78 Topics in Current Chemistry

Fortschritte der Chemischen Forschung

Biochemistry

Springer-Verlag
Berlin Heidelberg GmbH 1979

This series presents critical reviews of the present position and future trends in modern chemical research. It is addressed to all research and industrial chemists who wish to keep abreast of advances in their subject.

As a rule, contributions are specially commissioned. The editors and publishers will, however, always be pleased to receive suggestions and supplementary information. Papers are accepted for "Topics in Current Chemistry" in English.

ISBN 978-3-662-15446-5   ISBN 978-3-540-35376-8 (eBook)
DOI 10.1007/978-3-540-35376-8

Library of Congress Cataloging in Publication Data. Main entry under title: Biochemistry. (Topics in current chemistry; 78) Bibliography: p. Includes index. CONTENTS: I. Hasselbach, W. The sarcoplasmic calcium pump. A model of energy transduction in biological membranes. Krampitz, G. and Witt, W. Biochemical aspects of biomineralization. Nozaki, M. Oxygenases and dioxygenases. 1. Biological chemistry–Adresses, essays, lectures. I. Series. QD1.F58 vol. 78 [QP509] 540'.8s [574.1'92] 79-4657

© by Springer-Verlag Berlin Heidelberg 1979
Originally published by Springer-Verlag Berlin Heidelberg New York in 1979
Softcover reprint of the hardcover 1st edition 1979

Triltsch, Graphischer Betrieb, 8700 Wurzburg
2152/3140–543210

Contents

The Sarcoplasmic Calcium Pump

A Model of Energy Transduction in Biological Membranes

Wilhelm Hasselbach

Max-Planck-Institut für medizinische Forschung, Abteilung Physiologie, Jahnstraße 29,
6900 Heidelberg, Germany

Table of Contents

1 Introduction

Ion Pumps in Biological Membranes and Their Suitability for a Mechanistic Analysis

This contribution deals with the adenosine triphosphate-dependent active transport of calcium ions in the membranes of the sarcoplasmic reticulum. The sarcoplasmic reticulum is an intracellular membrane structure which represents a characteristic feature of fast contracting striated muscles. It is arranged as an elaborated system of tubules and cisternae between the myofibrils. The system's sole function concerns the regulation of muscle activity by releasing and removing calcium ions. Vesicular-shaped fragments of these membranes are preparations exceptional in simplicity for studying the transformation of chemical into osmotic energy and its reversal. The isolated membranes allow to measure in the same assay the transport of calcium ions against their electrochemical potential energized by ATP, the occurrence of transport intermediates and the synthesis of ATP driven by the electrochemical potential of the created calcium gradient[1−4].

Abbreviations

EDTA   ethylenediaminetetraacetic acid
EGTA   ethyleneglycol bis (2-aminoethyl)-N,N'-tetraacetic acid
CCCP   m-chlorocarbonyl-cyanide phenylhydrazone

1.1 Cellular Equilibration Processes Depending on Active Sodium Potassium Transport

Evidence for the necessity of an energy-dependent transport of solutes and water in biological systems began to emerge around 1940[5−7]. Although the term "la constance du milieu intérieur" denoting the ability of an organism to maintain a constant composition of its aqueous phase was already introduced by Claude Bernard 1878[8], specific mechanisms for its realization were not envisaged. In multicellular organisms the maintenance of the internal milieu is accomplished at two different levels.

1. Like the free living single cell, all cells of an organism are equipped with a great number of transport mechanisms. They operate in the cell membranes which separate the cytoplasm from the extracellular space surrounding the cells. These systems denoted as pumps enable the cells to balance their osmotic and ionic composition and to accumulate or eliminate various metabolites.

2. The composition of the extracellular compartment is kept constant by the cooperation of the cellular transport systems with various transport systems which are located in epithelial layers connecting the organism with its environment (Table 1). Examples are the epithelia of the intestine, the kidney and of various glands. It is their function to establish a relatively constant concentration of the respective solutes

3

Table 1. Cation distribution in red blood cell and muscle

| | Cell Water | | | Plasma |
| | Red Blood Cell | | Mammalian Skeletal | |
	Man (mM)	Dog	Muscle (mM)	
Na	20	135	16	150
K	135	10	152	4–5
Ca	10^{-4}	–	10^{-4}	3

in the extracellular space of the organisms. In the narrow sense, this is what Claude Bernard meant when he inaugurated the term "milieu intérieur". Its stability is an essential prerequisite for the function of all multicellular organisms. The complex trans-cellular transport systems accomplish absorption or secretion by using the same basic mechanisms as they operate in the single cell.

The most universal transport systems are those involved in the transport of the ubiquitous inorganic ions, sodium, potassium and calcium[1]. The sodium pump counteracts passive water movement across the cell membrane by removing sodium ions together with chloride or other anions from the cytoplasm to lower its content of osmotically active substances. In most cells, however, the elimination of sodium ions is connected with an accumulation of potassium ions[6]. For three sodium ions leaving the cell two potassium ions are taken up[9, 10]. The resulting concentration

Table 2. Ion fluxes in resting cells

		Skeletal muscle	Giant nerve fiber of the squid $p \cdot mol \cdot cm^{-2} \cdot s^{-1}$	Red blood cell
Sodium	Influx	3.5	40	0.05
	Efflux	3.5	40	0.05
Potassium	Influx	8.8	–	–
	Efflux	5.4	–	–
Calcium	Influx	0.1	–	0.0003
	Efflux	0.1	–	0.0003

If the ions move passively and there is no interaction between the ions the ratio of influx to efflux is given by the following relation: $\frac{\text{Influx}}{\text{Efflux}} = \frac{Na_0}{Na_i} \exp\left(\frac{V.F.}{R.T.}\right)$. The negative internal potential of 90 mV for muscle and nerve fibers and the high sodium activity outside would yield a flux ratio much greater than one. Consequently, an active transport mechanism must contribute to the efflux of sodium[9].

---

1 The transport of protons will not be considered in this context. There is no evidence that proton pumps are involved in the regulation of the cells' acid base balance. Proton gradients which are generated across the membranes of cell organells or bacteria by metabolic or light driven reactions are either used to drive energy requiring reaction directly, for example substrate movements, by co- or countertransport or they are used to synthesize ATP by a membrane-bound ATP synthetase.

potentials of sodium and potassium ions represent an energy store which is used by many cells to drive other transport systems like sugar or amino acid transport[9]. The activities of the sodium potassium transport systems related to the size of the surface of the particular cell exhibit great differences. Table 2 illustrates that the red cell is equipped with a very feeble sodium potassium pump while its activity in nerve tissue is extremely high as it is in kidney and brain tissue. The concentration potentials of sodium and potassium ions in combination with a selective permeability of the membranes for these ions can give rise to electrical potentials across the membranes. In most cells the concentration ratio of sodium and potassium ions does not exceed values of approximately 30 which in cells with a high potassium permeability gives rise to membrane potentials of approximately 100 mV. A great number of cellular activities is regulated by slow and rapid passive ion movements across the membranes induced by local permeability changes[11].

1.2 Regulation of Cellular Activity Depending on Active Calcium Transport

As compared to the importance of the sodium potassium pump for osmotic balance and excitability, a specific influence on cellular functions of the cytoplasmic concentrations of sodium and potassium ions per se, seem to be negligible. The ratio of the intracellular concentrations of both ions can vary in a wide range without noticeably affecting cell metabolism (Table 1). In contrast, calcium ions interfere specifically with a great variety of cellular functions at extremely low concentrations and changes in the concentration of ionized calcium are used for the regulation of numerous cellular activities. An essential prerequisite for the regulatory effect of calcium ions is their very low intracellular ion activity. In fact, in erythrocytes[12], in muscle[13, 14] and giant nerve fibers[15] where the activity of ionized calcium could be measured, it was found to be lower than 1 μM (Table 1). This concentration is more than 1000 fold smaller than that existing in the extracellular fluid. The effect on various functions of small changes of the intracellular calcium level could be ascertained by applying calcium chelating agents, calcium ionophores, calcium-sensitive dyes, and calcium-sensitive luminescing proteins. A more or less uncontrolled increase of the intracellular calcium level is induced by the extracellular application of calcium ionophores which by incorporation into the bilayer of the plasma membrane increase its permeability for calcium ions. The occurring increase of the intracellular calcium activity can be monitored either by calcium-sensitive dyes or by the protein aequorine which at low concentrations of calcium ions emits light. A controlled increase of the concentration of ionized calcium in muscle and nerve fiber has been achieved by perfusing or injecting solutions containing a mixture of calcium and EGTA as calcium buffers[13].

The contractile proteins of the muscle were the first biological structures for which the role of ionized calcium as regulating agent has been proven. Direct and indirect evidence has been furnished showing that muscles' rapid activation and inactivation cycle depends on the sudden release of calcium ions and their subsequent complete removal[16—18]. During activation the calcium activity of the myoplasm rises transiently from $\sim$ 0.05 to 1 μM. In fast contracting muscles this change takes place during 10 to 100 ms. The quantities of calcium ions which have to be set free to cause this sudden start of chemical and mechanical activity are quite large because the contractile proteins need for saturation the considerable quantity of approximately 0.2 μmol calcium/ml Therefore, calcium stores with a quite high capacity on the one hand, and a powerful calcium removing system on the other hand, must be present in the intimate vicinity of the protein filaments of the contractile apparatus. Calcium does not interact with the same protein constituents of the contractile proteins in all muscles. In some muscles, calcium ions combine with myosin, the main constituent of the contractile machinery, while in others a regulatory protein system is the target for calcium ions[16].

The structures involved in calcium release and removal are the membraneous membrane network of the sarcoplasmic reticulum. The mechanism of calcium release is virtually unknown. No reasonable concept has emerged from numerous experiments dealing with the problem[17]. In contrast, considerable evidence concerning the mechanism of calcium withdrawal has been obtained showing that the rapid removal of calcium ions is brought about by a very active ATP-driven calcium transport system in the sarcoplasmic reticulum membranes[18]. Apart from this intracellular membrane system responsible for the fast activation and inactivation of the muscle, muscle cells like all other cells are equipped with calcium transporting systems in their plasma membranes which, on the long run, maintain the low internal calcium level. The transport activity of these systems is much lower than that of the sarcoplasmic reticulum membranes. This can be so, because the rates with which calcium ions invade the muscle fiber across the plasma membranes are quite low due to the relative impermeability of these membranes for calcium ions. One of the transport systems in the plasma membrane acts by exchanging internal calcium ions for external sodium ions[15, 19]. The energy for calcium extrusion is thus furnished by the sodium potential. Since the sodium ion gradient is relatively shallow while the calcium ion gradient is very steep, sufficient energy is available only from the sodium potential when three sodium ions are exchanged for one calcium ion. For the removal of the entered sodium ions, the sodium potassium pump must become active. Thus, this calcium transport is indirectly driven by the sodium potassium pump. In addition to this system some cells are equipped with a calcium transport mechanism *sui generis,* directly coupled with the metabolism of the cell[20].

1.3 Identification of the Energy Source and the Energy Transforming Reaction of Active Ion Movement

The origin of the energy which the cell has to provide for ion movement was controversial until 1960. At that time conclusive evidence was presented that the sodium potassium transport system is fueled by energy-rich phosphate compounds[21−25]. This progress was mainly made possible by the application of methods which allow to manipulate reproducibly the internal composition of the giant nerve fiber of the squid axon[21] as well as of the red blood cell[23] with respect to ions and energy furnishing substrates. While the participation of energy-rich phosphate compounds was firmly established in these experiments, the view that ATP might be the immediate energy donor could not be ascertained. Unambiguous proof for this concept was first obtained for the sarcoplasmic calcium transport. The active calcium transport performed by isolated sarcoplasmic reticulum vesicles is most effectively driven by ATP while phosphate compounds, like creatine phosphate or arginine phosphate which are the most prominent energy-rich phosphate compounds in the muscle cells cannot be used as energy donors[2, 26]. Other phosphate compounds which can be consumed by the membranes (comp. 3.1.5.) are not present in sufficient concentrations in the muscle cell. The difficulty to assure the energy source for the sodium potassium pump is an inherent problem for all sodium potassium transporting membranes. Preparations like the red blood cell or the squid axon which allow to measure easily ion movement are not well suited for studying the chemical events connected with transport processes. It is hardly possible to supply the transport system in these cells with a defined energy donor because the cells contain various phosphoryl transferring enzymes, so that the added phosphate compound is not necessarily identical with the compound used by the transport system. Furthermore, it is quite difficult to keep the ion concentration in the internal space of the cells under permanent control. Finally, the proteins which are specifically involved in ion translocation can be isolated from these preparations only in very small quantities which makes it impossible to investigate physical or chemical changes of these proteins as they may occur during transport.

Some of the discussed problems could be approached after Skou (cf.[27]) had demonstrated that in membrane fragments isolated from kidney and brain, an ATP cleaving system was present which needs for optimal activity the simultaneous presence of sodium and potassium ions at concentrations as they are required for the activity of sodium potassium transport. These preparations allow to study various chemical events such as hydrolysis of ATP or the phosphorylation of the membrane proteins by ATP, as they depend on the concentrations of sodium and potassium ions. Yet in these preparations the different chemical events cannot be correlated to ion transport. The fragmented membranes never form tightly sealed vesicles suitable to study ion translocation. The attempts to incorporate membrane fragments or solubilized and purified membrane proteins into ion-impermeable liposomes could only partially overcome these difficulties[28, 29]. Hence, it is not possible to study sodium potassium transport and its chemistry simultaneously. Most important for the development of a coherent scheme of the sodium potassium transport was the finding of a specific inhibitor of sodium potassium transport, ouabain[30]. It suppresses sodium potassium transport in living cells as well as the sodium potassium ATPase of fragmented membranes with the same effectiveness[22, 25].

In contrast to the plasma membrane fragments the fragments of the sarcoplasmic reticulum membranes form vesicles which retain the accumulated ions. What is likewise important, these membranes mostly maintain their original sidedness in spite of the complete destruction of the natural arrangement of the reticulum during isolation. The fact that the transport substrate as well as the energy yielding substrate must be delivered to the external surface of the sarcoplasmic reticulum vesicles offers great experimental advantages. Furthermore, since the surface density of the transport protein is very high in these membranes a correspondingly high transport activity results. Hence, the isolated sarcoplasmic reticulum vesicles possess nearly all the properties a preparation should have to enable one to investigate simultaneously both, ion movement and the concomitant chemical reactions, energy yielding or conserving as well as the occurrence of reaction intermediates in membrane constituents[1-4].

Until recently, active ion transport was considered mainly as a process in which metabolic energy was unidirectionally consumed for the performance of osmotic work. Yet, a new aspect emerged when P. Mitchell[31] proposed that proton gradients created by metabolic reactions may be used to drive the synthesis of ATP in mitochondria, chloroplasts or bacteria by reversing ATP hydrolysis. In the following, it has been demonstrated that ATP synthesis cannot only be driven by the potential of proton[32] but also by the sodium potassium[10] or the calcium ion potentials[2]. However, the latter two systems differ, at least, in one essential respect from the proton translocating ATPase. The translocation of sodium and potassium ions as well as that of calcium ions is correlated with the formation of different phosphoryl compounds as reaction intermediates in the membranes. That such intermediates could not be found in the proton translocating ATPase poses considerable difficulties to envisage a mechanism by which inorganic phosphate is incorporated into ADP and, therefore, has led to the proposition of various speculative concepts[33].

2 Characterization of the Sarcoplasmic Reticulum Membranes and Their Calcium Pump

2.1 Identification

The most prominent and long known structure of the striated muscles are their contractile elements arranged in myofibrils which are characterized by an alternating band pattern. The transverse bands are constructed of areas of thick and thin protein filaments. They comprise the contractile machinery (cf.[34, 35]). Another structure, also specific for the striated skeletal muscle, yet of much greater delicacy, is located in the sarcoplasm filling the space between the myofibrils. This structure was first seen by Koellicker, Cajal and Retius around 1890 (cf.[36]). Ten years later, Veratti[37] published the results of a most comprehensive study based on light microscopy of silver-stained muscle sections. He drew and described the interfibrillar structure as a reticulum network of thin fibers arranged in different patterns depending on the particular muscle. In the following, mention of these structures has almost disappeared until they were rediscovered by Bennett and Porter[38] in the electron microscope. The fibrous network was identified as being composed of two elaborated membrane systems, the system of transverse tubules which are narrow invaginations of the plasma membrane and the sarcoplasmic reticulum which constitute an intracellular complex network of membranous tubules and cisternae (Fig. 1a)[17]. The transverse tubules have been identified as the structures along which the action potential is conducted inwards.

The identification of the sarcoplasmic reticulum membranes as the intracellular calcium transport system involved in the regulation of muscle activity took considerable time and occurred at first quite fortuitously. Kielley and Meyerhof[40], not aware of the existence of the sarcoplasmic reticulum, characterized it biochemically as an

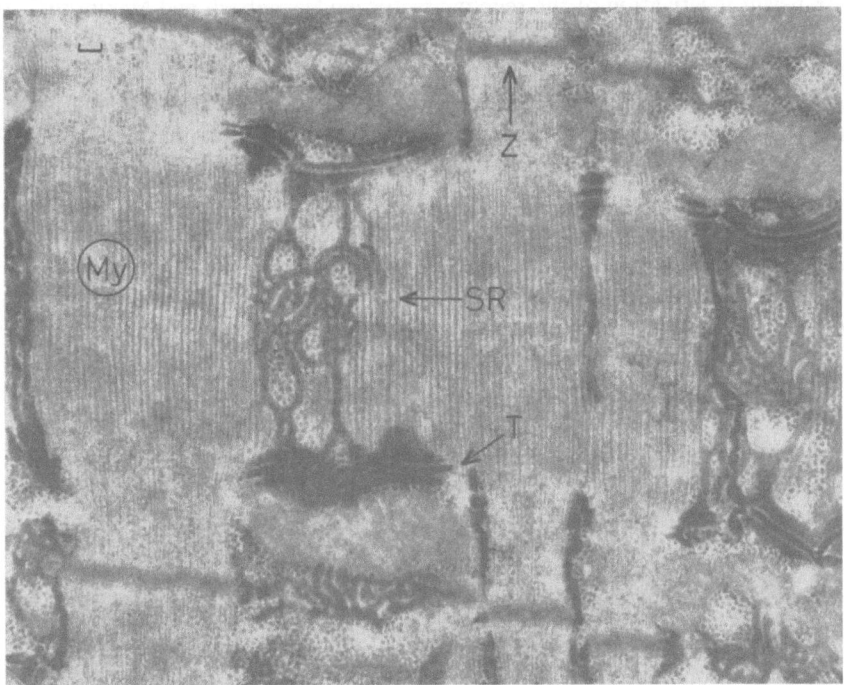

Fig. 1a. Electronmicrograph of a muscle section form the diaphragm of the mouse showing sarcoplasmic reticulum and transverse tubules contrasted with lanthanum[39]. SR, sarcoplasmic reticulum; T, transverse tubule; My, myofibril; Z, Z line

ATP hydrolyzing enzyme which they found in aqueous extracts of homogenized muscles. The enzyme was defined by the pronounced activation produced by magnesium ions and by the severe inhibition caused by physiological concentration of calcium ions. Furthermore, the authors observed that the enzyme was inactivated by the action of phospholipase C. thus demonstrating its lipoproteinaceous nature. The very same enzyme was found by Ulbrecht[41] as a contamination in different preparations of contractile proteins and it was shown that in addition to ATP hydrolysis the enzyme catalyzes a fast exchange of phosphate between ATP and ADP. Another and most important activity in the aqueous muscle extract, later attributed to the fragmented reticular membranes, was discovered by Marsh[42]. Marsh had found that a factor present in aqueous muscle extracts could inhibit the ATPase of the contractile proteins and could either prevent their ATP-induced contraction or induce relaxation of contracted preparations when ATP was present. Therefore, the factor was called "relaxing factor". Furthermore, Marsh demonstrated that relatively low concentrations of calcium ions could abolish the effect of the relaxing factor. The particulate nature of the relaxing factor was revealed when Portzehl[43] succeeded in removing it from aqueous muscle extracts by centrifugation. At the same time Ebashi[44] concentrated the factor by ammonium sulfate precipitation. Shortly later, the presence of vesicular-shaped membranes was detected in the precipitated material[45]. The observation that under certain conditions calcium ions did not produce a permanent but

9

only a transient inactivation of the relaxing factor led Hasselbach and Makinose[26] to study the interaction of the vesicular fragments with calcium ions. It was demonstrated that the vesicular membrane fragments are able to store considerable quantities of calcium ions by an ATP supported process. Surprisingly, the calcium accumulation was tremendously enhanced by the presence of calcium precipitating anions like oxalate which were used to protect the material against the inactivating effect of the ubiquitous calcium contaminations. In the presence of oxalate, calcium oxalate precipitates were formed inside the vesicles and these precipitates could be seen in the electron microscope as electron dense material (Fig. 1b)[1, 46]. This ATP-dependent calcium storage was used to verify the view that the calcium storing vesicles originate from the sarcoplasmic reticulum membranes. Hasselbach[47] and Costantin et al.[48] reported the appearance of massive calcium precipitates inside the muscle fiber where the cisternal elements of the reticulum are located (Fig. 1c). Since the external membranes of the muscle fiber are very impermeable for calcium ions, they can reach the sarcoplasmic reticulum only after the permeability barrier of the plasma membrane has been removed. This was achieved either by stripping it off[48] or by making it permeable by a glycerol treatment[47].

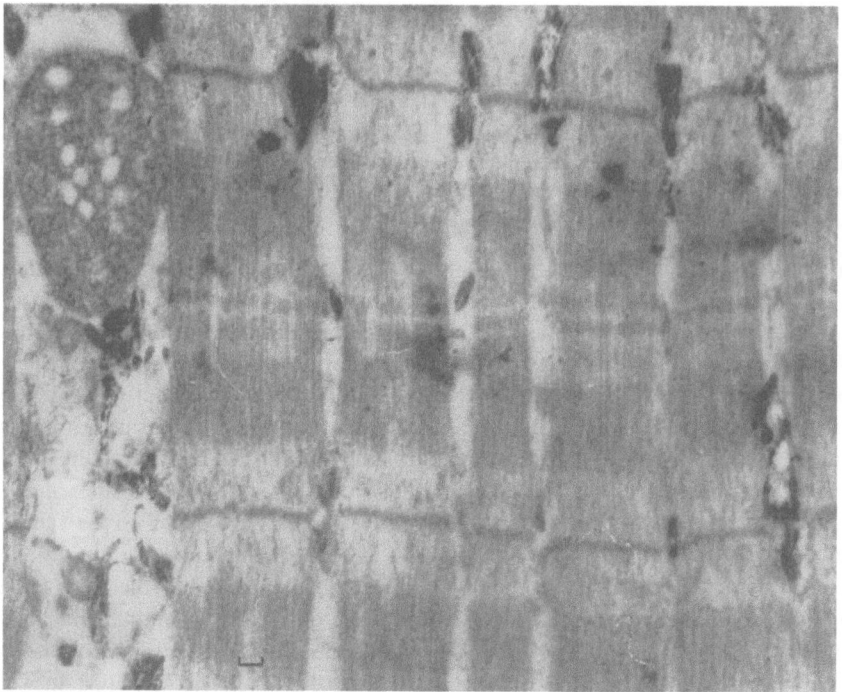

Fig. 1 b. Electronmicrograph of a muscle section from the sartorius muscle of the frog showing calcium oxalate crystals in the cisternal enlargements of the sarcoplasmic reticulum[47]. During the preparation the cisternae were disconnected from the tubular structures of the sarcoplasmic reticulum. In the frog muscle the T tubules and the cisternae of the sarcoplasmic reticulum are localized near the Z lines while in the diaphragm of the mouse the two structures follow the border line between the thick and the thin filaments of the myofibrils (My)

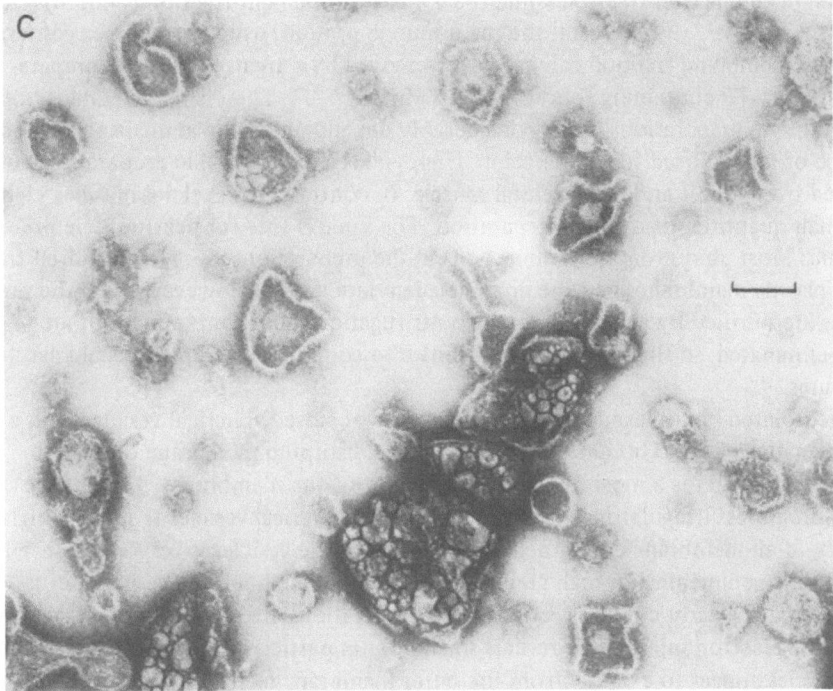

Fig. 1 c. Electronmicrograph of isolated sarcoplasmic reticulum membranes loaded with calcium oxalate[46]. The preparation was negatively stained with potassium phosphotungstate. Size and shape of the calcium oxalate precipitates differ considerably. Fine particles which are present on the surface of most vesicles are clearly visible. In all pictures the length of the bars measures 0.1 μm

2.2 Isolation and Purification of Fragmented Sarcoplasmic Reticulum Membranes

Fragments of the sarcoplasmic reticulum membranes have been isolated from a great variety of muscles. Isolation and characterization of the membranes is facilitated by their characteristic properties, namely by their ability to store large quantities of calcium and by their ATPase activity which is stimulated by low concentrations of calcium ions[49] Isolation starts with muscle homogenization in salt or sucrose containing solutions. Unbroken muscle fibers, bundles of myofibrils, single myofibrils, nuclei, plasma membrane fragments and mitochondria are removed from the homogenate by centrifugation at medium gravitational forces. The fragments of the sarcoplasmic membranes which remain in the supernatant are subsequently sedimented at high speed centrifugation. This crude preparation is usually fractioned by centrifugation in sucrose containing solutions of different densities or through a sucrose gradient. The membranes which possess the highest calcium-dependent ATPase activity and the highest calcium storing capacity are usually found in the fraction with a density of 1.13 to 1.17 g/ml[50]. The less active heavier fractions differ in some respect from the lighter one. Drugs like caffeine or rhyanodine interfere with calcium accumulation of the heavier fraction while the activity of the light fraction is not affected by the substances[51, 52]. Myosin, the main muscle protein, which sometimes contaminates the membrane fraction can easily be removed by a treatment of the preparation with an ATP containing solution of 0.6 M KCl[53, 54]. The yield and stability of the membrane preparations differ considerably depending on the animal species and the type of muscle used for preparation. High yields and very stable preparations are obtained from rabbit and frog skeletal muscle. In contrast, red skeletal muscles yield only small quantities of a labile preparation. The same is true for heart muscle preparations. Most ubiquitous contaminations in the preparations are mitochondrial fragments, plasma membranes and the enzyme adenylate kinase. However, when the preparations are purified by sucrose gradient centrifugation these contaminations are largely eliminated, so that they do not give rise to complications in the usual analytic procedures.

The isolated sarcoplasmic membranes consist of closed spherical vesicles with a mean diameter of 80−100 nm. The ability of the disrupted membrane tubules to form closed vesicles is a most remarkable property of the membranes. The change from more or less irregular membrane fragments to spherical vesicles requires a high mobility of all membrane constituents The shape of the vesicles is very stable, even when during sedimentation high gravitational forces are applied the vesicular cross section remains nearly circular[55]. When the preparations are positively or negatively stained the electron microscope reveals small 4,0 nm particles in the membranes[56−58]. The particles appear to extrude from the outer membrane leaflet in negatively stained preparations. Particles also cover the outer leaflet of freeze fractured preparations (Fig. 2). These particles appear to be considerably larger and less frequent than those

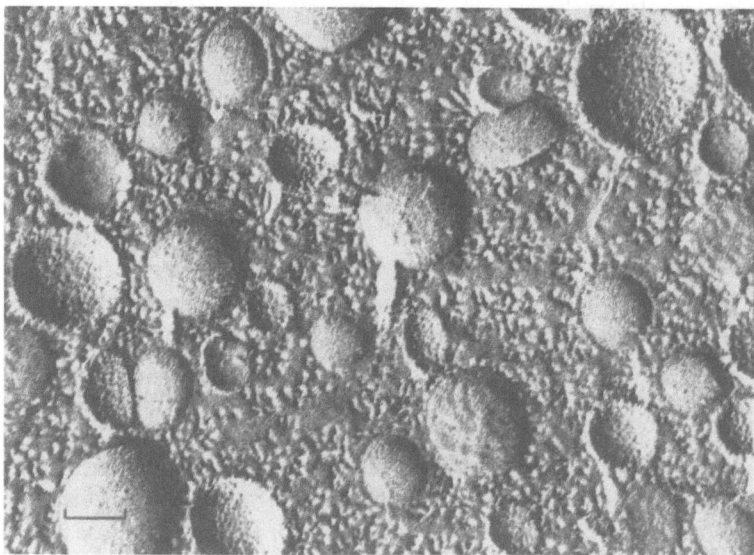

Fig. 2a. Electronmicrograph of a freeze fractured preparation of isolated sarcoplasmic reticulum vesicles. 9 nm membrane particles are clearly visible on the concave cytoplasmic fracture faces of the vesicular membranes

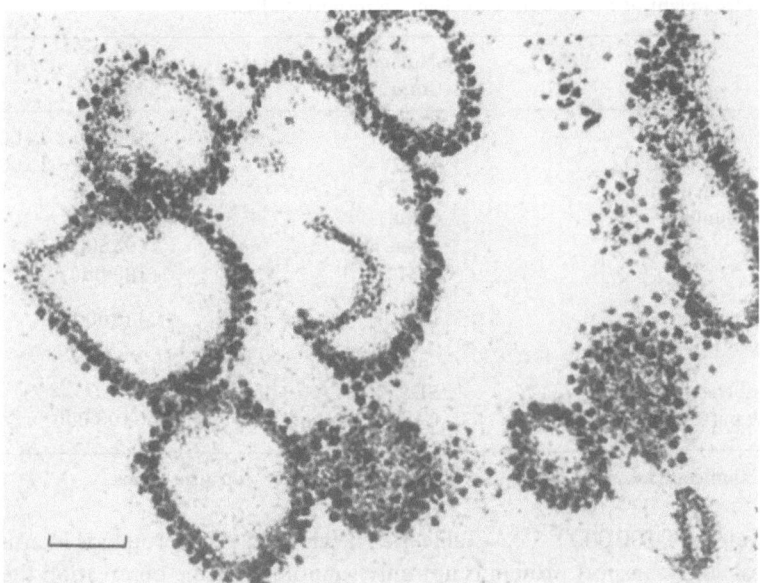

Fig. 2b. Electronmicrograph of a sectioned preparation of isolated sarcoplasmic reticulum vesicles decorated with the electron dense thiol reagent mercuri phenylazoferritin. Note the absence of ferritin particles on the internal surface of the membrane fragment in the center of the picture[50]

seen in stained preparations[59–61]. It has, therefore, been suggested that the larger particles may consist of aggregates of the smaller ones[62].

2.3 Constituents of the Sarcoplasmic Reticulum Membranes

2.3.1 Proteins of the Sarcoplasmic Reticulum Membranes

2.3.1.1. The Transport Protein. The high degree of specialization of the sarco-plasmic membranes together with kinetic properties of their calcium transport system (comp. 3.1.4.) suggested that a relatively high percentage of the proteins in the membranes might be involved in calcium transport. A first approach to estimate the amount and the molecular weight of the calcium transporting polypeptide was based on the observation that calcium transport could specifically be blocked by substitution of only a few thiol groups in the membrane protein with thiol reagents like N-ethyl-maleimide[63]. When the membranes labelled with ^{14}C N-ethyl-maleimide were solubilized by succinylation and separated by gel filtration in sodium dodecylsulfate containing solutions the labelled material was eluted from the sepharose 4B column at a volume characteristic for a molecular weight of ~ 100000.–. Very similar values for the molecular weight of the eluted material were found by ultracentrifugation using both s and D, or s and η values as well as sedimentation equilibrium results[64]. Finally, sodium dodecylsulfate gel electrophoresis has established that 70–80 % of the protein matrix of purified membrane preparations consists of the protein with

13

Table 3. Molecular weight of the solubilized calcium transport ATPase

	Molecular weight from	M
Succinylated membranes dissolved in 0.25 M Na_2HPO_4 0.2% SDS fractionated on a Sepharose 4 B column [64]	$(s° D°)$	102000 ± 14000
	$(s° \eta)$	98000 ± 15000
	$(D° \eta)$	112000
	(Equ.)	130000
	(Osm.)	93600 ± 5500
	(Gel filtr.)	100000
Deoxycholate solubilized and delipidated ATPase [74]	(Equ.)	115000
Membranes dissolved in SDS containing buffers [65–69]	SDS Gel electrophores.	~100000

$s°$ = Sedimentation coefficient, $D°$ = diffusion coefficient, η = intrinsic viscosity.

a molecular weight of 100000[65–68] (Table 3). The view that this protein is identical with the calcium transport protein is not only supported by the observation that the protein labelled with thiol reagents migrates together with the main protein component but also by the fact that it contains all radioactivity which is incorporated into the membranes when they are incubated with ATP labelled in the γ-position with ^{32}P under conditions of calcium transport (comp. 5.2.1.).

The isolation of the pure and enzymatically functioning main membrane protein was accomplished by MacLennan[69] who succeeded to obtain the enzyme at low yields as a lipoprotein complex by applying deoxycholate as detergent. In recent years MacLennan's procedure was modified in many laboratories and 80 % of the enzyme present in the membrane can be isolated as a pure lipoprotein preparation. Its protein-lipid ratio is not very different from that of the native vesicles. Instead of deoxycholate other mild detergents like cholate[67], lysolecithin[70] or Triton X-100[71] can be used for the removal of the contaminating accessorial proteins from the native membranes. Yet, care has to be taken in choosing appropriate quantities of detergents in relation to the amount of the membranes. When detergents like cholate or deoxycholate are used in quantities sufficient to remove the phospholipids completely from the protein by the formation of mixed micells, the structure of the protein changes and its calcium-dependent ATPase becomes inactive[72]. The resulting deoxycholate protein complex contains approximately 0.3 mg deoxycholate per mg protein[73]. The molecular weight of the protein moiety of the complex has been found to be 115000 with a stock radius of 5 nm and a friction ratio of 1.5[74]. The detergents lysolecithin or Triton X-100 oxythylene glycol displace the natural lipids like cholate. In contrast, however, the protein retains most of its characteristic enzymatic activities. Evidently, these detergents can substitute, at least to a certain extent, the natural lipids[70, 75], comp. 4.2.

2.3.1.2. Accessorial Proteins. In sodium dodecylsulfate gel electropherograms a few smaller protein components comprising less than 20 % of the total membrane protein migrate ahead of the main protein. Different migration patterns have been reported indicating that these components might be present in different quantities in different preparations and/or that the mobility in the gel may depend on the used separation system. Since variable patterns of these protein components observed in the same preparation system could not be related to calcium transport activity, the minor protein constituents seem not to be essential for calcium transport. As minor protein components of the sarcoplasmic reticulum membranes, proteins designed as calcium precipitating protein[71], calsequestrin, high affinity calcium binding protein and lipoprotein[4, 69, 76–78] were described (Fig. 3). The protein constituent which can most easily be isolated in good yields is the calcium precipitating protein. After its removal from the membrane by applying low concentrations of Triton X-100 or

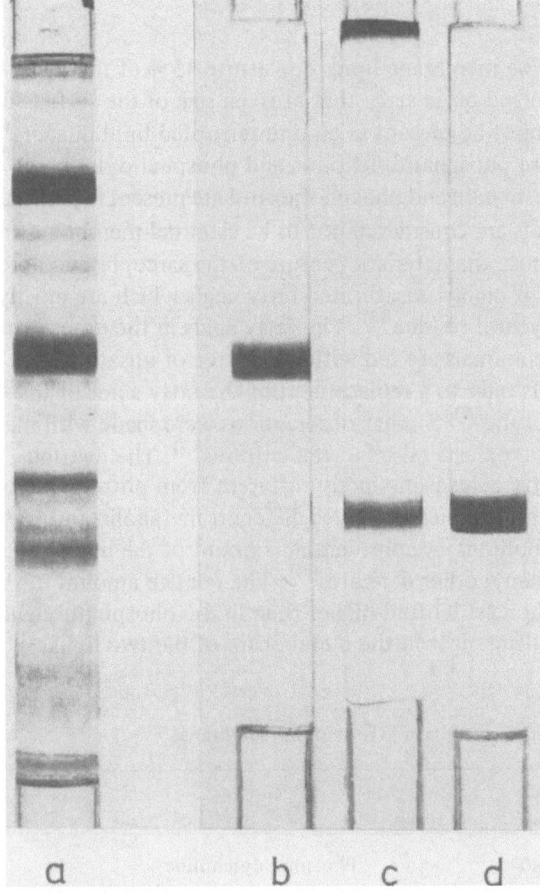

Fig. 3. Electrophoretic separation of the proteins of the sarcoplasmic reticulum membranes in sodium dodecylsulfate containing solution on polyacrylamid gels.
a) Crude preparation with fast moving accessorial proteins and a slow moving component which is presumably an oligomer of the transport protein.
b) Purified transport protein.
c) Purified calcium precipitating protein.
d) Serum albumin as reference

deoxy-cholate it can be precipitated by calcium ions. After a cycle of dissolution by calcium removal and precipitation by calcium addition, the protein can be considered as pure. It has relatively high calcium binding capacity but a quite low calcium affinity[79]. For its molecular weight determined by sodium dodecylsulfate gel electrophoresis values were reported between 55000 and 70000. Most likely, the calcium precipitating protein is identical with calsequestrin for which MacLennan[76] has reported a molecular weight of 44 000. The large discrepancies of the molecular weight are presumably due to an abnormal behaviour of the protein under the different experimental conditions applied for electrophoretic separation in sodium dodecylsulfate containing buffers.

2.3.2 Membrane Lipids

2.3.2.1. Lipid Composition. The membrane lipids constitute 35 % of the membrane matrix[80-82]. It was recognized quite early that at given size of the surface of the membranes the lipid phase cannot be present as an uninterrupted lipid bilayer[56]. The predominant phospholipids are phosphatidylcholine and phosphatidylethanolamine. Phosphatidylserine, sphingomyelin and phosphoinositol are present in considerably smaller amounts. Neutral lipids are considered not to be essential membrane constituents[80, 82-84] (Table 4). A most characteristic feature of the sarcoplasmic lipid is the presence of a great fraction of highly unsaturated fatty acids which are mostly located in the a-position of the glycerol residue[85]. The fatty acids in the a-position remain unsaturated even when the animals are fed with a diet free of unsaturated fatty acids (Table 5). The diet leads only to a replacement of the fatty acids of the linoleic type by those of the oleic type[86]. Similar observations were made with membrane preparations isolated from myoplasts raised in cell cultures[87]. The distribution pattern of the unsaturated fatty acids is distinctly different from phosphatidylcholine and phosphatidylethanolamine which indicates different metabolic pathways for the two most abundant phospholipids. A considerable amount of the main phospholipids is present as their a,β-alkenyl ether derivates[88]. The relative amount in the phosphoethanolamine fraction is at least tenfold higher than in the phosphatidylcholine fraction which again stresses differences in the metabolism of the two lipid fractions.

Table 4. Proteins and phospholipids of the sarcoplasmic reticulum membranes

	%		%
Transport protein	70–80	Phosphatidylcholine	64
Calcium precipitating protein (Calsequestrin)	15	Phosphatidylethanolamine	20
		Phosphatidylinositol	10
High affinity calcium binding protein	?	Phosphatidylserine	3
		Sphingomyelin	3
Lipoprotein	?		
Glycoprotein	?		
67, 68, 77, 92)		82, 83)	

Table 5. Fatty acid composition of lipids of sarcoplasmic
vesicles from control and fatty acid deficient rats[86]

Fatty acid	Amount of fatty acid in:	
	Controls	Fatty acid deficient
	%	%
16:0	37.8 ± 4.3	30.2 ± 5.0
16:1ω7	3.1 ± 0.7	8.0 ± 2.1
18:0	7.1 ± 1.5	4.1 ± 1.4
18:1ω9	18.0 ± 2.3	29.4 ± 4.4
18:2ω6	11.6 ± 1.8	1.4 ± 0.9
20:3ω9	2.4 ± 1.4	16.3 ± 3.2
20:4ω6	7.2 ± 1.5	2.3 ± 1.0
22:6ω3	5.8 ± 1.6	1.2 ± 0.6

2.3.2.2. Properties of the Lipid Matrix of the Membranes. In spite of the relatively small lipid area between the domains of the membrane proteins a fast lateral diffusion in the plane of the membrane takes place. At 37°C a diffusion constant of
$6.10^{-8} cm^2/s$ has been calculated for spin-labelled phosphatidylcholine molecules[89].
This diffusion constant would allow a phospholipid molecule to move with a rate of
several microns/s. Such a fast lateral movement indicates a considerably fluidity of
the lipid phase which is based on the high degree of unsaturation of the membrane
lipids. As in other membranes the phospholipids seem to be asymmetrically arranged.
Phosphatidylethanolamine and phosphatidylserine appear to be exclusively located
in the cytoplasmic leaflet while phosphatidylcholine is predominant in the leaflet facing the intravesicular space[90−92]. As to the lateral organization of the phospholipids,
it must be considered that the large fraction of the membrane lipids is in intimate
contact with the transport protein. An estimate of the number of phospholipid molecules which might be in close contact with the protein is based on the observation
that when the phospholipids are removed from the membrane the protein starts to
loose its enzymatic activity when the number of phospholipid molecules per transport protein becomes smaller than $30^{93, 94)}$. 30 phospholipid molecules can form a
single bilayer belt around the hydrophobic portion of the transport ATPase if it has
a diameter of ∿20 A. Spin-labelled phospholipid molecules present in this belt seem
to be more rigid and immobilized than in the lipid bilayer distant from the protein.
However, it remains ambiguous if this layer of phospholipid molecules can be considered as a separate phase since no thermal transition phenomena have been observed
supporting their existence[95].

2.3.3 Arrangement of Lipids and Proteins in Sarcoplasmic Reticulum Membranes

The low lipid-protein ratio of 0.5 together with the size of the sarcoplasmic vesicles
implies that only approximately 30% of their membranes can be occupied by a
regular lipid bilayer structure. Consequently, a large fraction of the membrane protein must interrupt the lipid bilayer and reach through it. The fact that only one polypeptide chain constitutes the structural unit of the calcium transport protein strong-

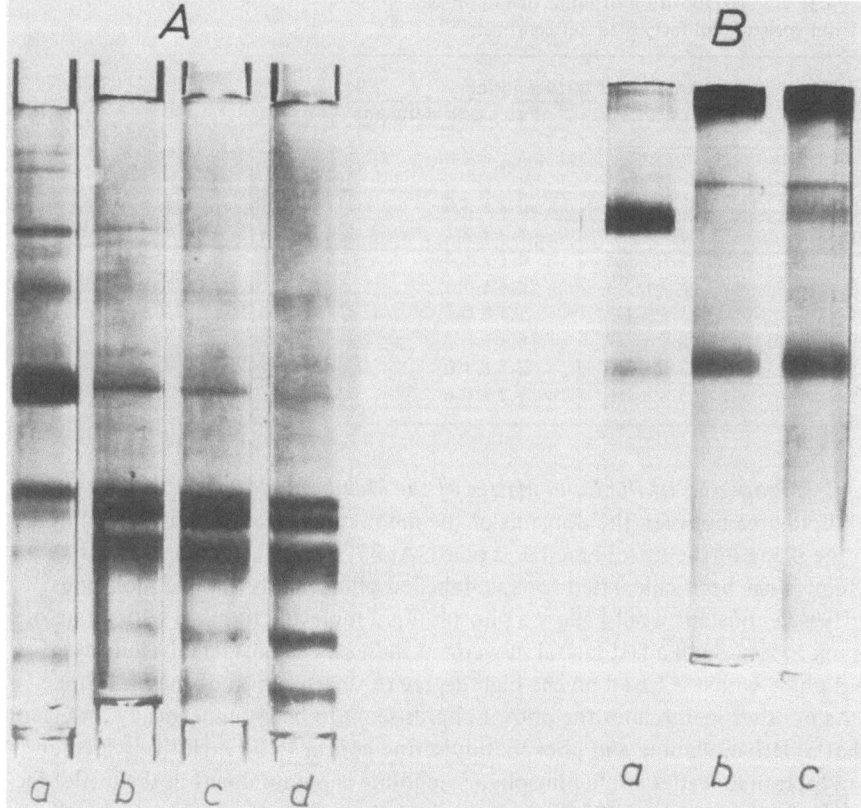

Fig. 4. Separation of trypsin digested and crosslinked sarcoplasmic reticulum protein[96].
A. Preparations digested for a) 0′; b) 2′; c) 5′; and d) 30′ with trypsin 0.005 mg per mg of
   membrane protein.
B. Preparations after crosslinking the surface proteins with dextran aminophenyldiazotate
   a) Control preparation not crosslinked,
   b) preparation was shortly digested with trypsin and subsequently treated with dextran
      aminophenyldiazotate,
   c) normal preparation treated with dextran aminophenyldiacotate

ly suggests that the transport molecule itself is in contact with both sites of the membrane. This arrangement is in agreement with the finding that no material with a
molecular weight of 100 000 enters the polyacrylamide gel after the protein at the
outer surface of the sarcoplasmic reticulum vesicles has been crosslinked by dextranaminophenyldiazotate and solubilized by sodium dodecylsulfate[90] (Fig. 4). The
same arrangement of the transport protein in the membrane infers from experiments
in which closed and open vesicles were labelled with the fluorescing dye Fluoresca-
min (Fluram, Hoffmann-La Roche) under conditions which do not result in a disruption of the membranes[96, 97]. The results suggest that from the total surface of the
protein accessible from both sites of the membrane 70 % are in contact with the cytoplasmic, and 30 % with the vesicular space. This distribution is in agreement with the
result of X-ray studies performed with oriented multilayers of the membranes which

18

were obtained by controlled unisodiametric drying[98, 99]. An asymmetric arrangement of the transport protein also infers from the accessibility of its thiol groups. The electron dense thiol reagent mercuriazobenzeneferritin decorates only the cytoplasmic leaflet of nonvesicular membrane fragments[56] (Fig. 2b). In the protein itself, structural domains must be present which are arranged perpendicular to the plane of the membrane. These domains give rise to a high intrinsic birefringence, the axis of which is perpendicular to the normal of the membrane plane[100]. The view that we are dealing with an asymmetric molecule is further supported by the finding that so solubilized transport protein appears as spindle-shaped particles in the electron microscope[72]. The section of the transport protein which extrudes from the cytoplasmic leaflet is easily and specifically attacked by trypsin[90, 101−104]. Low concentrations of trypsin cleave the molecule in two fragments of similar molecular weights of 45 000 and 50 000, without any loss in its ATPase activity. The fragments embedded in the membrane are even able to perform calcium transport. As one should expect, sections of the tryptic fragments are in contact with the cytoplasmic surface. This has been confirmed by labelling the membranes with antibodies raised against tryptic fragments[105, 106]. During prolonged digestion trypsin cleaves readily the larger protein fragment while the smaller one seems to be more resistant. This has been taken as indication that the smaller fragment might comprise the more hydrophobic section of the molecule by which it is anchored in the lipid bilayer. However, such an arrangement is in conflict with the observation that all tryptic fragments embedded in the vesicular membrane are crosslinked by externally applied dextranaminophenyldiazotate[90] (Fig 4b).

The arrangement of the accessorial proteins in the membranes is unknown or a matter of conjecture[101, 107]. The localization of calsequestrin has been tried to ascertain by different approaches. The fact that it can be stained by dyes which react only with the external membrane surface argues for its location in the membrane itself[96, 101]. On the other hand, the quite small extent of labelling observed under very mild conditions has been interpreted in favour of an internal location of calsequestrin[92]. An intravesicular location is also suggested by the observation that dextranaminophenyldiazotate crosslinks the ATPase while calsequestrin remains unaffected[90]. The resistance of calsequestrin against trypsin digestion and its inaccessibility for antibodies after external application also seems to support an internal location of the protein[90, 102, 105, 106]. The observed labelling pattern can be expected if a small segment of calsequestrin is anchored in the membrane while the larger part of the molecules reaches into the internal space of the vesicles[96, 97].

3 Energy Dependent Calcium Translocation

3.1 ATP Driven Calcium Influx

3.1.1 Calcium Storage in the Absence of Calcium Precipitating Anions

Native sarcoplasmic reticulum vesicles rapidly remove calcium from solutions containing ATP, magnesium ions, and ionized calcium at concentrations between 1 and

10 μM. The maximum amount of calcium ions which is stored by the vesicles is limited to approximately 150 nmol calcium/mg protein[108-111]. The minimal level to which the concentration of ionized calcium in the external fluid can be reduced is in the range of 1 to 5 nM[49, 112]. This ATP-dependent calcium removal occurs as long as the sarcoplasmic reticulum membranes exist as tightly sealed vesicular structures. Neither membrane fragments obtained by sonication nor the purified ATPase are able to sequester calcium ions. On the other hand, calcium ions stored by closed vesicles are completely released when the structure is destroyed or if calcium ionophores are incorporated into the lipid matrix of the membranes. It can, therefore, be excluded that calcium storage is brought about by an ATP-supported calcium binding mechanism. The specificity of the ATP-dependent storing mechanism is not restricted to calcium ions, strontium ions are accumulated in a quite similar manner[111, 113]. The time course of calcium uptake has been measured by separation of the calcium loaded vesicles from the residual calcium in the solution by filtration through filters of small pore size[114] and by flow-dialysis technique[117] (Fig. 5). The time resolution of both methods, however, is insufficient owing to the fast rate of uptake and the relatively small storing capacity of the vesicles. A sufficient time resolution can be obtained with optical methods which require the application of calcium indicator dyes[115, 116]. Under the assumption that the calcium ions remain soluble inside the vesicles the total internal calcium concentration would reach approximately 10 mM. This figure is obtained with a value of 5 μl for the volume which is occupied by one mg of vesicular protein[117]. Since at a calcium concentration of 2–5 mM broken membranes bind passively, i.e. in the absence of ATP, the same quantity of calcium, namely 100 to 150 nmol/mg, as it is bound by closed vesicles in the presence of ATP in solutions containing 1–10 μM calcium ions, part of the stored calcium should be bound to internal binding sites[118, 119]. Calcium binding proteins, phospholipids and the calcium

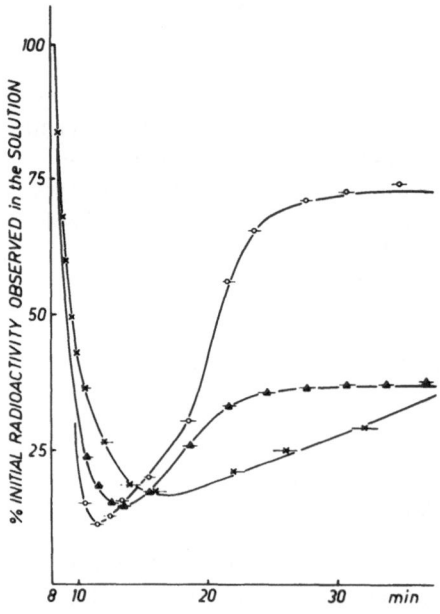

Fig. 5. Calcium uptake of sarcoplasmic reticulum vesicles supported by different substrates followed by calcium release after substrate depletion[113]. The time courses of calcium uptake and calcium release were measured by flow dialysis. The reaction chamber contained 0.4 mg protein/ml, 5 mM MgCl$_2$, 50 mM Tris-maleate pH 7.45, 0.1 M KCl, 0.05 mM CaCl$_2$. ○–○ ITP; ▲–▲ ATP; x–x acetyl phosphate

transport ATPase itself have been considered as possible candidates of these sites. Since however, their affinity for calcium is quite low, a large fraction of the stored calcium will remain unbound. A free internal calcium concentration of 2–5 mM may be a fair guess.

This concentration is at least 100 to 1000 times higher than the concentration of ionized calcium existing in the external solution when the ATP-supported calcium uptake reaches saturation. The high concentration gradient created by the ATP-depending calcium transport mechanism discipates either when the energy yielding substrate is removed or exhausted or if the energy yielding reaction is blocked[111–113] (Fig. 4). This indicates that the formation and the maintenance of the concentration gradient requires permanent energy supply. The rate with which calcium leaks through the membranes after they are disconnected from their energy source is quite low. Even if the calcium concentration in the external medium is kept extremely low by the presence of EGTA, the rate of calcium release does not exceed 10 to 20 nmol/mg and min at room temperature[120]. This rate is approximately 100 times slower than the rate with which calcium ions are taken up actively. The relatively slow passive calcium efflux through the sarcoplasmic membranes is an essential prerequisite for the efficiency of the accumulation process.

3.1.2 Calcium Storage in the Presence of Calcium Precipitating Anions

In contrast to the relatively small amount of calcium ions stored by the sarcoplasmic vesicles, 100 fold higher quantities are taken up when the incubation medium is supplemented with calcium precipitating anions like oxalate, phosphate or pyrophosphate[26, 108, 121]. Self-evidently, the concentration of ionized calcium in the incubation medium has to be adjusted to low values not allowing spontaneous precipitation of the calcium salts in the assay medium. This can most easily be achieved by the addition of the calcium sequestering agent EGTA which enables one to adjust the concentration of ionized calcium between 10 nM and 1 μM in media of neutral pH. Under such conditions, calcium and oxalate or calcium and phosphate are found to be taken up in quasi stoichiometrical amounts, and calcium precipitates are formed inside the vesicles (Table 6). The uptake of calcium always exceeds that of the anions by a small quantity[49, 108, 121]. That the uptake of the anions never surpasses that of calcium excludes that calcium accumulation might be brought about by an active ac-

Table 6. Stoichiometric relationship between the storage of calcium and oxalate or phosphate in sarcoplasmic reticulum vesicles

Stored ions	No precipitating anions present			1 mM oxalate present			10 mM secondary phosphate present		
	(n)	μmol/mg prot.	S.E.	(n)	μmol/mg prot.	S.E.	(n)	μmol/mg prot.	S.E.
Ca^{2+}	(6)	0.102	±0.008	(13)	2.66	±0.14	(5)	2.014	±0.062
Anion		–		(8)	2.46	±0.18	(4)	1.743	±0.17

The assay medium contained: $CaCl_2$ 0.5 mM, EGTA 0.7 mM, Mg ATP 2 or 5 mM, pH 7.0[121].

cumulation of the anions followed by calcium precipitation when due to the increase of the anion concentration inside the vesicles the solubility product is reached[49]. In the presence of oxalate or phosphate containing media calcium uptake proceeds until the vesicles are "full", if more calcium is added than the vesicles can store, or if less calcium is present, calcium uptake continues until the calcium concentration in the solution has been reduced to a limiting value of approximately 1—5 nM.

3.1.3 Energy Requirement of Calcium Transport

As a consequence of the precipitation of calcium ions inside the vesicles, the product of the ion activities of calcium and the respective anions in the internal vesicular space is fixed by the solubility product of the precipitating salts (L). Since in the external solution changes of the ion activities of calcium (Ca_o) and of the precipitating anions (A_o) can be followed experimentally, the energy requirement for calcium uptake can be determined at every moment. The energy requirement is given by the expression:

$$E = R \cdot T \cdot \ln \frac{V_i}{V_e} + R \cdot T \cdot \ln \frac{L}{Ca_0 \cdot Ox_0}$$

V_i = rate of calcium influx; V_e = rate of calcium efflux; L = solubility product of calcium oxalate.

 The first term in the equation gives the energy which is needed to maintain net calcium uptake. When net calcium uptake ceases, the energy requirement is given by the second term alone which represents the reversible osmotic work[2]. The uncertainty caused by the use of the solubility product for determining the energy requirement concerns mainly the dependence of its value on the ionic strength of the medium. However, this dependence becomes relevant only at values of the ionic strength much higher than those possibly present inside the vesicles[122]. When the uptake of calcium is not limited by the calcium storing capacity of the vesicles, calcium uptake proceeds until the activity ratio $L/Ca_o \cdot Ox_o$ approaches values between 2000 and 10000. These values correspond to an energy requirement of 16.000 to 20.000 KJmol calcium taken up (Table 7). The cessation of net calcium uptake when an amount of calcium is offered which exceeds the storing capacity of the vesicles is difficult to understand. It is not caused by a visible destruction of the vesicular membranes.

 Because of the large quantity of calcium which can be stored by the vesicles in the presence of oxalate or phosphate the time period of calcium uptake becomes sufficiently long to follow the kinetics of calcium uptake accurately. In the presence of oxalate or phosphate calcium uptake cannot only be terminated by removing the calcium loaded vesicles from the solution by filtration but also by precipitation with mercuri salts at slightly acid pH[49]. Another important advantage acquired by the addition of calcium precipitating agents to the assay media is the fact that the measured initial rate of calcium uptake becomes practically identical with the rate of calcium influx. As the result of calcium precipitation inside the vesicles the internal calcium activity remains low and constant, therefore, passive calcium efflux can be neglected as compared to the rapidly proceeding uptake of calcium. The question

Table 7. Performance of the sarcoplasmic reticulum calcium pump driven by ATP or UTP[2]

	Performance of the calcium pump		ATP	UTP
During net uptake	Rates (20 °C) pmol mg$^{-1} \cdot$ min^{-1}	Ca influx	1.7	0.46
		Ca efflux	0.10	0.025
	Concentration ratios	$\dfrac{L}{Ca_0 Ox_0}$	25	25
	Energy requirement	KJ/mol	~16	~16
During steady state	Rates (20 °C) pmol mg$^{-1} \cdot$ min^{-1}		0.8	0.3
	Concentration ratios	$\dfrac{L}{Ca_0 Ox_0}$	2500	2100
	Energy requirement	KJ/mol	~20	~20

whether calcium uptake in the presence of oxalate proceeds as fast as in its absence has recently been answered by following the initial rate of calcium uptake in the absence of precipitating anions by fast recording optical methods and the rate of calcium uptake was found to be identical under both conditions[123].

3.1.4 Coupling Between ATP Splitting and Calcium Transport

The uptake of calcium is usually initiated by the addition of calcium ions to the otherwise complete assay medium. Before calcium is added, ATP is split with a low rate by most sarcoplasmic reticulum preparations. This ATPase which is active in the absence of calcium ions has been named "basic ATPase".

When, on addition of calcium ions calcium uptake starts, ATP splitting instantly rises 3 to 10 fold, provided that the concentration of ionized calcium remains between 1 and 10 μM. This calcium induced ATPase has been called "extra ATPase"[26] (Fig. 6). Calcium uptake and ATPase activity are not fortuitously simultaneous events[114]. Their causal linkage is revealed by the invariance of the coupling ratio. Under various conditions, two calcium ions are taken up by the sarcoplasmic vesicles for each molecule of ATP split[108]. This constant coupling ratio is exhibited by the native sarcoplasmic vesicles, although the absolute rate of calcium uptake may differ by more than a factor of 10. At very low rates of calcium uptake deviations from the coupling ratio have been reported[49, 124]. Yet, one has to consider that under these conditions the increment of extra splitting is small and its determination as the difference between the total and the basic splitting becomes uncertain. In the presence of a high concentration of monovalent salts[125] which suppress the ATPase only moderately and at temperatures

23

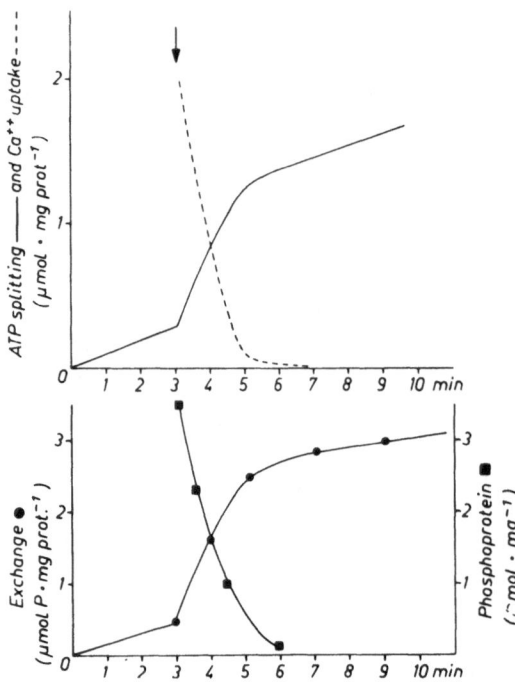

Fig. 6. ATP cleavage, ATP–ADP exchange and phosphoprotein formation during calcium uptake. The assay medium contained 7 mM MgCl$_2$, 5 mM ATP, 2 mM ADP, 5 mM potassium oxalate, 40 mM KCl, 0.2 mM EGTA, 10 mM histidine pH 7.0, T = 20 °C. ^{14}C labeled ADP and γ ^{32}P labeled ATP were used for the exchange and the phosphorylation experiments, respectively. Calcium uptake is illustrated as the removal of calcium from the assay medium. The arrow indicates the addition of 0.1 μmol calcium chloride · ml^{-1} to the medium containing 0.05 mg protein · ml^{-1}. Note, basic splitting and basic exchange prior to calcium addition. The rapidly formed phosphoprotein disappears when the calcium concentration declines

above 37°C[95], a true decline of the coupling ratio seems to occur. The observation that under most conditions a constant coupling ratio is observed, is very difficult to reconcile with the view that the calcium uptake is the result of an ATP-dependent influx and a passive efflux of calcium ions through leaks in the membranes. The above mentioned low calcium permeability alone is not sufficient to explain the invariant coupling ratio. The rate of calcium efflux can vary considerably without affecting the coupling ratio or the maximal concentration ratio reached after the cessation of net calcium uptake. The latter value should even be more sensitive to variation of the calcium efflux than the coupling ratio. Therefore, one has to envisage some kind of coupling between the energy providing reaction and both calcium influx and calcium efflux.

3.1.5 The Specificity of the Energy Yielding Reaction

The sarcoplasmic transport system can be fueled in addition to ATP by a great number of phosphate compounds which differ considerably in their chemical nature. Not only the natural nucleoside triphosphates[126] but also para-nitrophenylphosphate[127], acetyl phosphate[128] or carbarmyl phosphate[129] can drive calcium transport. While there are considerable differences between the rates with which calcium transport proceeds with the different substrates, they are all used with the same coupling ratio of two and the pump can establish similar maximal concentration ratios (Fig. 7) (Table 8). Evidently, stoichiometry and transport energetics are substrate-independent properties of the pump. The kinetics of the transport, however, is very much affected by the substrate. This substrate specificity is not very well understood. It has been

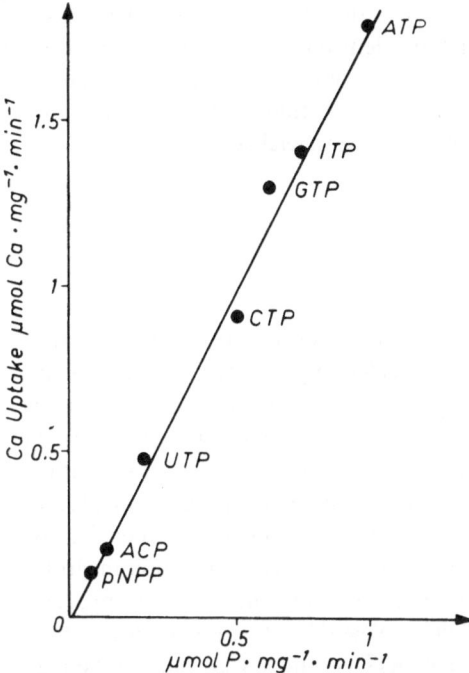

Fig. 7. Correlation between the rate of calcium uptake and the rate of calcium activated liberation of phosphate from various energy yielding substrates [138]

Table 8. Residual calcium levels and maximal concentration ratios accomplished by the sarcoplasmic reticulum calcium pump fueled with different substrates [55, 126]

Energy donor	Residual ionized calcium nM	Concentration ratio $L/Ca_0 \cdot Ox_0$
ATP	1.6	2800
ITP	1.5	2980
GTP	2.0	2370
CTP	1.9	2200
UTP	3.0	2100
ACP	1.0	4700

The level of ionized calcium is calculated from the measured residual, radioactive calcium and the stability constant of Ca EGTA = $5 \cdot 10^6 \, M^{-1}$ (pH 7.0).

proposed that the substrates interact at two sites with the transport molecule [130]. At one site, they enter the reaction cycle as energy yielding substrates and at the other site, they act as more or less effective activators. The fact that substances like acetyl phosphate or para-nitrophenyl phosphate can drive calcium transport, is of special interest because these substrates were thought to be hydrolyzed by the sodium potassium transport system in a terminal reaction step occurring after ion translocation [10]. In the absence of calcium ions all substrates are hydrolyzed with approximately the same rate by a basic magnesium-dependent phosphatase. The affinity of the nucleotides for this enzyme species is considerably lower than for the calcium-dependent enzyme[108]. Furthermore, the magnesium-dependent phosphatase is not inhibited by thiol reagents. The question is unsettled whether the basic phosphatase is an enzyme per se which might perhaps reside in membrane invaginations of the plasma membrane present as impurities or if we are dealing with a special state of the calcium activated enzyme[131].

3.2 Calcium Efflux Coupled ATP Synthesis

The view that not only calcium influx but also calcium efflux might be connected with the enzymatic activity of the sarcoplasmic calcium pump could in fact be confirmed. In the course of permeability studies performed with calcium loaded vesicles Barlogie et al.[132] discovered that a rapid release of calcium ions through the calcium transport molecule could be induced. It was shown that the slow passive calcium release was accelerated 10—50 fold when the release medium was supplemented with phosphate and ADP. The fact that the acceleration of calcium efflux needs the presence of the splitting products of ATP strongly suggested that the reaction might be the reversal of the ATP consuming calcium uptake. In fact, a net synthesis of ATP could be demonstrated to occur during calcium release[133—137]. Furthermore, the stoichiometry of the calcium efflux-dependent ATP synthesis was found to be iden-

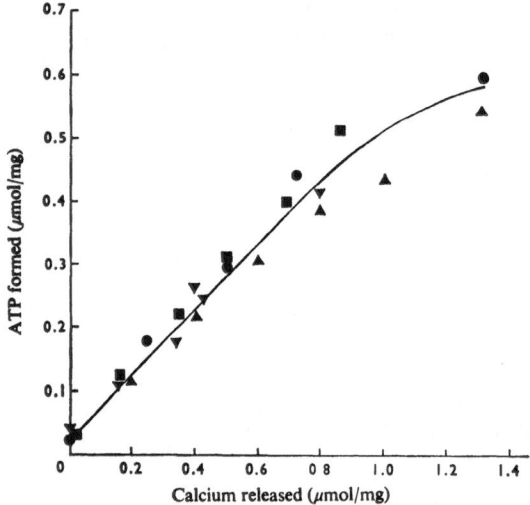

Fig. 8. Correlation between calcium release and ATP formation[138]. The sarcoplasmic reticulum vesicles were loaded with calcium phosphate in media containing 0.5—2.0 mM acetyl phosphate, 10 mM $MgCl_2$, 5 mM potassium phosphate, 0.5 mg protein/ml pH 7.0, 0.1 M glucose and 0.05 mg/ml hexokinase at 30 °C. To obtain different calcium loads, different amounts of calcium chloride were added. During the incubation period of 30′ ~ 99% of the added calcium were taken up. Calcium efflux and ATP synthesis were started by adding an equal volume of a solution containing 2 mM EGTA, 10 mM $MgCl_2$, 5 mM ^{32}P inorganic phosphate and 2 mM ADP. The radioactive glucose-6-phosphate was determined after inorganic phosphate had been removed by extraction of its molybdate complex with isobutanol benzene

tical with that of the active calcium accumulation, i. e., for two calcium ions released from the vesicles one molecule of ATP was synthesized (Fig. 8). This correlation is valid when the total amount of ATP synthesized is compaired with the amount of calcium released by ADP from vesicles loaded with 0.2 to 1 μM calcium per mg protein[135, 138]. Evidently, all calcium released by ADP from the vesicles contributes to ATP formation. This is the case when the conditions were chosen in such a way that the reaction can unidirectionally run to completion. An increase of the calcium concentration in the external medium must be prevented by the addition of excess EGTA, and the level of ATP must be kept low by the transfer of its terminal phosphate to glucose. Under these conditions, phosphate incorporation into glucose via ATP can proceed until the ratio between internal to external calcium approaches the value of 10. At this low ratio practically all calcium is released from the vesicles.

Thus, in sarcoplasmic reticulum membranes a conversion of osmotic into chemical energy can most easily be demonstrated, and since the concentrations of all reactants are known at every moment, the energetics of the process is well defined. This reversibility is described by the following over all reaction (Table 9).

Table 9. Energy requirement for calcium gradient driven ATP synthesis

$$2\,Ca_0 + ATP \rightleftharpoons 2\,Ca_i + ADP + P_i$$

$$\Delta G = \Delta G_0' - RT \cdot \ln \frac{ATP \cdot Ca_0^2}{ADP \cdot P \cdot Ca_i^2} \tag{1}$$

$$= \Delta G_0' - RT \cdot \ln \frac{ATP}{ADP \cdot P_i} - 2\,RT \cdot \ln \frac{Ca_0}{Ca_i} \tag{2}$$

$$= (\Delta G_0' - \Delta G_0'') - RT \cdot \ln \frac{GP}{G \cdot P} - 2\,RT \cdot \ln \frac{Ca_0}{Ca_i} \tag{3}$$

$$16.7 + 5.9 \cdot \log \frac{10^{-4}}{10^{-1} \cdot 5 \cdot 10^{-3}} \leqslant 11.7 \cdot \log \frac{Ca_i}{Ca_0}$$

$$12.5 \leqslant 11.7 \cdot \log \frac{Ca_i}{Ca_0} \qquad \qquad \Delta G_0' = -31.0$$
$$\Delta G_0'' = -14.5$$

$$10 \leqslant \frac{Ca_i}{Ca_0}$$

In Eq. (3) the hexokinase reaction ATP + glucose = ADP + glucose-6-phosphate with a standard free energy of $\Delta G_0'' = -14.5$ kJ/mol is introduced. $\Delta G_0'$, standard free energy for ATP hydrolysis.

3.3 Inhibition of Energy Transduction

3.3.1 Reversible Inhibition

While the sodium potassium ATPase and the ATPase of the mitochondria or chloroplasts can quite selectively be blocked by inhibitors which are effective at low concentrations, such specific inhibitors are not known for the calcium transport system. Neither ouabain which selectively blocks the sodium potassium ATPase[10, 25] nor 2-4-dinitrophenol or CCCP which interfere with ATP synthesis in mitochondria, affect the calcium transport system of the sarcoplasmic reticulum[115, 135]. For the calcium-dependent ATPase and calcium uptake only inhibitors of relatively low specificity have been found. They are pharmacological agents which produce local anesthesia, β-receptor blockade or tranquilizing effects[81, 139]. Calcium efflux and ATP synthesis are depressed likewise by these agents. As to the mechanism of inhibition it has been shown that their inhibiting effect is reduced when the calcium concentration is increased and, therefore, a competition between the inhibitors and calcium ions has been proposed[140]. On the other hand, the inhibiting effect of the agents is considerably enhanced in preparations whose phospholipids were hydrolyzed by phospholipase A_2[85]. The latter finding favours the view that the effect of the drugs is mediated by the lipids in the membrane.

Reversible inhibition of the sarcoplasmic reticulum calcium transport is also produced by reagents which perturbate the interaction of the transport complex with ATP or other substrates. Chaotropic anions like NO_3^-, ClO_4^-, CCl_3COO^- are such agents[141]. They inhibit calcium transport and calcium-dependent ATP splitting to about the same extent when present at concentrations > 0.1 M. The anions simultaneously enhance passive calcium permeability of the membranes but the resulting calcium efflux is too low to account for the diminished calcium accumulation. It could rather be demonstrated that the simultaneous inhibition of ATP splitting and calcium accumulation is brought about by a competition of the chaotropic anions for ATP at its binding site on the transport protein. Most likely, the anions due to their low degree of solvation occupy hydrophobic pockets of the transport protein. In other enzymes such pockets are thought to be the sites for nucleotide binding[142]. A completely different type of agents which reversibly affect the calcium transport and calcium-dependent ATP splitting are substances like dimethylsulfoxid and ethyleneglycol[143]. They do not interfere with ATP binding as chaotropic anions do. These reagents obviously affect later steps in the reaction sequence (comp. 5.2.1.).

A most interesting reversible inhibitor of the calcium transport system is arsenate[144]. No effect, whatsoever, is exerted by arsenate on the calcium-dependent ATPase and calcium accumulation. Arsenate affects solely the system when it functions as ATP synthetase. Apart from calcium ionophores[115, 145] arsenate is the only agent which effectively uncouples calcium efflux driven ATP synthesis. When it is added to an assay containing calcium loaded vesicles, EGTA and ionized magnesium, a fast calcium release is induced (Fig. 9). This arsenate induced calcium release does neither require the presence of ADP nor of phosphate. These reactants rather uncompetetively inhibit the effect of arsenate. It is most likely brought about by the formation of an arsenylated intermediate which by spontaneous hydrolysis allows calcium efflux.

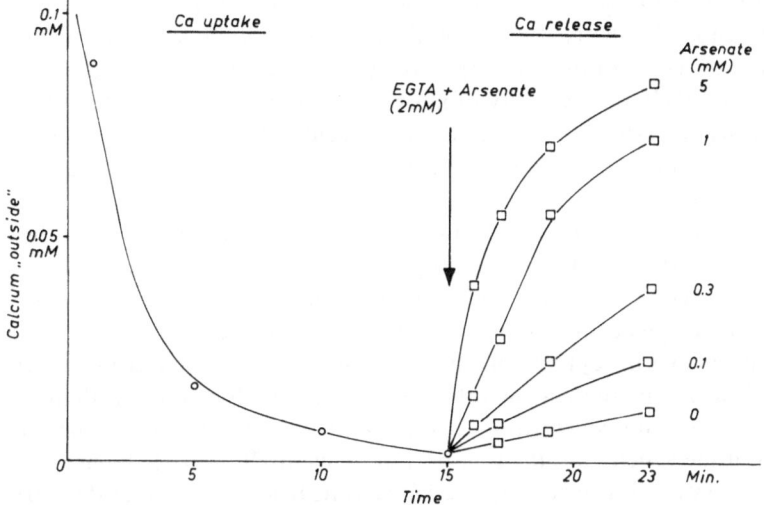

Fig. 9. The arsenate induced calcium release from calcium loaded vesicles[144]. The vesicles were loaded with calcium phosphate in an assay containing 5 mM MgCl₂, 2 mM acetyl phosphate, 20 mM inorganic phosphate, 0.1 mM CaCl₂ and 0.1 mg · ml⁻¹ vesicular protein. Calcium release was induced by the addition of arsenate and 2 mM EGTA. Note that the addition of EGTA alone only causes a very slowly proceeding calcium release

3.3.2 Irreversible Inhibition

It was very early recognized that the calcium transport and the calcium-dependent ATPase could simultaneously be blocked by thiol reagents[26]. In contrast to various other thiol containing enzymes the activities of the sarcoplasmic reticulum membranes cannot be restored when the blocking agents are removed.

In spite of several attempts to determine the total number of thiol residues in the membrane protein its value remained controversal for a long time. When the thiol groups in the native enzyme are titrated, only 8–10 groups in one transport molecule are accessible for the substituting reagents. Additional 4–6 groups become substituted when the membranes are solubilized by sodium dodecylsulfate[146–148]. A total of 22 thiol groups per molecule of purified enzyme has been found recently after the protein was treated with sodium dodecylsulfate and reducing agents[149]. Obviously, in the native enzyme a variable number of the thiol groups seemed to form SS-bridges.

There is general agreement that the enzyme activity declines linearly with the number of substituted thiol groups. Yet, the total number of thiol residues which must be blocked to suppress activity completely varies between 4 and 20, depending on the preparation and on the applied conditions[146, 150–152]. Interestingly, nucleoside tri- and diphosphates modify the time course of the reactions by which the thiol groups are blocked by a variety of reagents. The pseudo-first order rate of substitution becomes slower in the presence of nucleotides depending on their concentrations. In contrast to the nucleotides, energy donors like acetyl phosphate are completely ineffective[55]. We are obviously dealing with a change in the reactivity of the thiol groups induced by nucleotide binding. This change does not take place in a section of the molecule whose flexibility depends on the presence of lipids since lipid deprivation does not interfere with the protective effect of nucleotides. Nucleotide binding diminishes most likely the reactivity of most thiol groups accessible to the reagents and does not protect a particular group at the active center[146, 152]. The dependence of the protective effect on the nucleotide concentrations has been used to determine their affinity to the enzyme[64, 148, 153].

Calcium ions also seem to interfere with the reactivity of the thiol residues. The rate with which a few thiol groups react with the fluorescing molecule mercuri-cyste-indansylate is enhanced by low and reduced by high concentration of calcium ions[154]. We must assume that a great fraction of the thiol groups is located in the part of the molecule extruding from the cytoplasmic surface of the sarcoplasmic reticulum vesicles. The latter becomes more than any other membrane densely covered with mercuri-phenylazoferritin which is an electron-dense SH reagent that can neither penetrate the membranes of the vesicles nor the transport molecule itself[56]. The absence of thiol groups at the internal surface of the vesicles results from the fact that the ferritin compound becomes never attached to the internal membrane leaflet of open vesicular fragments (Fig. 2b).

4 Functional Aspects of Lipid Protein Interaction

4.1 Passive Calcium Permeation

The lipids in the sarcoplasmic reticulum membrane are essential for the formation of tightly sealed vesicles from the fragmented membrane network as well as for energy-dependent calcium translocation. As mentioned in section 3.1.1., the rate of passive calcium efflux from native vesicles does not exceed 20 nmol/mg and min at room temperature which corresponds to a calcium flux of 0.1 pmol per second and square centimeter. It is the same slow rate as observed for the calcium flux through the plasma membranes of the resting muscle (Table 2). But, as compared to the calcium flux through membranes of liposomes prepared from lipids isolated from sarcoplasmic membranes, native vesicular membranes must be considered to be quite permeable for calcium ions[155]. One may suggest that calcium evades from the native vesicles either through the transport protein per se or through defects in the lipid bilayer. As to the mechanism of passive calcium passage, the interaction between charged ion species seems to be of little importance: the calcium permeability is scarcely affected by the ionic strength of the medium[120, 125]. In contrast, however, low concentrations of chaotropic anions which interact with the transport protein specifically accelerate calcium release considerably[141].

The contribution of the lipids to the low permeability of the sarcoplasmic vesicles for calcium is revealed by the tremendous increase in the rate of calcium efflux which occurs when the lipid phase is modified. The following modifications proved to be effective: treatment of the sarcoplasmic reticulum vesicles with low concentrations of ether[156], splitting of a small fraction of the phospholipids by phospholipase A_2

or phospholipase C[85], incorporation of amphiphilic agents like Triton X-100, cholate, deoxycholate, oleate or lysolecithin into the vesicular membranes. To abolish the permeability barrier completely it suffices to split approximately 10 % of the phospholipid moiety[85] or to incorporate 0.05 or 0.1 mg cholate/mg vesicular protein[157]. Once the vesicles have been made calcium permeable, the calcium pump is unable to remove calcium from the medium. Consequently, the ATPase remains permanently active. We are not dealing with a true activation of the ATPase by this perturbation of the lipid matrix. The leaks in the membrane prevent the inactivation of the ATPase as it occurs by calcium removal from the medium or by raising the internal calcium concentration when tightly sealed vesicles are examined. The produced increase of the permeability cannot or only very incompletely be annulled by removing the amphiphiles from the membranes. The preparation from which the amphiphiles have been removed by dialysis, gel filtration or ion exchange are characterized by a permanently active ATPase. This indicates that the ability of the vesicles to store actively calcium ions is restored, if at all, only very incompletely. Apart from the discussed alteration of the lipid matrix of the membranes more specific permeability changes can be produced by the incorporation of lipophilic calcium ionophores like X 537 A (Hoffmann-La Roche) or A 23187 (Lilly)[115, 145] (Fig. 10). The rate of the occurring calcium release is considerably smaller than that found on bimolecular membranes of phosphatidylcholine for the effect of the ionophore A 23187[158]. The effect of the ionophores seems to be irreversible, presumably due to their high affinity for the membranes.

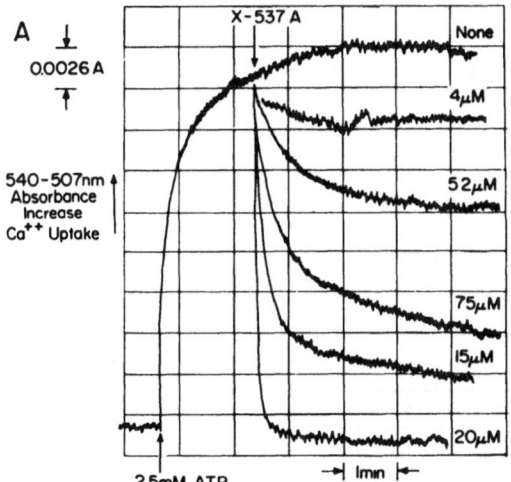

Fig. 10. The effect of different concentrations of the ionophore X 537 A on calcium release, by sarcoplasmic reticulum vesicles[115]. The reaction mixture contained 20 mM Tris-maleate pH 6.8, 50 mM KCl, 10 mM $MgCl_2$, 0.1 mM $CaCl_2$, 0.1 mM murexide and 0.27 mg protein/ml. Calcium uptake and release were followed by monitoring the changes in the absorbance undergone by murexide. The measurements were performed with a filter dual wave length (540−507 nm) double beam spectrophotometer

4.2 Degradation and Restitution of the Calcium-Dependent ATPase and Calcium Transport

The involvement of lipid membrane constituents in the interaction of ATP with the protein moiety of the calcium transport system emerges if one compares the reaction of ATP with membrane preparations whose lipid matrix has been removed or modified to varying degrees. Lipid depletion can be achieved by treating membrane preparations with excessive amounts of detergents like deoxycholate or Triton X-100 or by digestion with phospholipases. Vesicular lipids can gradually be removed by applying deoxycholate at appropriate concentrations and protein detergent ratios. Approximately 50 % of the membrane lipids can be removed without any reduction of the activity of the calcium-dependent ATPase[93, 94, 159]. Yet, if the number of phospholipid molecules per enzyme molecule become smaller than 30, the enzymatic activity starts to decline and disappears at a molar ratio of phospholipid to enzyme of 15. Various unsuccessful attempts were made to reconstitute enzymatic activity of completely delipidated preparations by readdition of different lipid components[55, 72].

Partial delipidation as it can be performed by treating the vesicles with phospholipase C or phospholipase A_2 in combination with albumin also leads to inactivation of the calcium-dependent ATPase and abolishes self-evidently calcium uptake[85]. If some residual enzymatic activity persists, it appears only at temperatures above 32°C. The preparations which have lost approximately 70–80 % of their normal phospholipid content easily regain full calcium-dependent ATPase activity. If a higher percentage of the phospholipids is removed with cholate, restitution of the enzyme activity is delicate and remains incomplete[160]. It is long known that reactivation does not need the complete diacyl phospholipid molecule. Simple single chain lipid components like fatty acids or lysophosphatidylcholine compounds can most effectively restore enzymatic activity[85, 161, 162]. Optimal ATPase activity is achieved when about 60 to 80 fatty acid residues are absorbed by the protein molecule[161]. The lipid com-

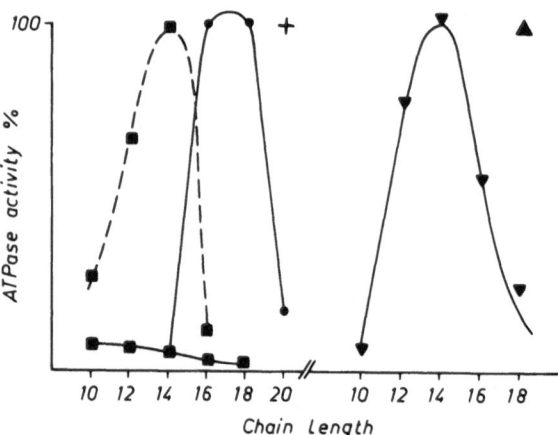

Fig. 11. Reactivation of calcium dependent ATPase activity of lipid deprived sarcoplasmic reticulum membranes by single chain lipid compounds. ■- - - -■ Saturated fatty acids: 0.5– 1.2 mg/mg protein, T = 32 °C; T = 20 °C. ■ —— ■ 100% = 2.5 μmol P · mg⁻¹ · min⁻¹. • —— • Unsaturated fatty acids with one *cis* double bond, and + arachidonic acid: 0.2 mg/mg protein, 100% = 1.1 μmol P · mg⁻¹. min⁻¹, T = 20 °C.▼ - - - -▼ Deoxy lysophosphatidylcholine with saturated fatty acid residues: 0.2–1.0 mg/mg protein, and ▲ oleyl deoxy lysophosphatidylcholine: 0.2 mg/mg protein; 100% = 1.1 μmol P · mg⁻¹ · min⁻¹, T = 20 °C

pounds with the shortest hydrocarbon chain which produce significant activations of the enzyme are laurylic acid and lauryl-lysophosphatidylcholine[163]. The activating effects of the saturated fatty acids and the lyso compounds with saturated fatty acids display sharp optima when the hydro-carbon chain is elongated (Fig. 11). Activation declines sharply when myristylic acid is replaced by palmitylic acid. Hence, laurylic and myristylic residues are the only saturated hydrocarbon chain which can produce substantial activation if present either as fatty acids or as lyso compounds. The essential difference between the activation effect of saturated fatty acids and lyso-phosphatidylcholine compounds with saturated fatty acid residues is the fact that the latter produce activation already at 20 °C, while the fatty acids need temperatures above 30°C. The activity of the reconstituted ATPase with C-16 and C-18 phosphatidylcholine reaches only 30—40 % of the optimum produced by C-14 lyso-phosphatidylcholine. Such a sharp decline of the activating effect of both fatty acids and the lyso compounds as the result of chain elongation does not take place when the chains contain cis double bonds, at least one, in position 9 to 10. This observation may be important for the understanding of the specific nature of lipoprotein interaction. Evidently, saturated fatty acids or their derivates can provide conditions allowing the ATPase molecule to reach its active conformation only if its fatty acid chain is longer than 10 carbon atoms and shorter than 16 carbon atoms. The ineffectiveness of the long chain components is most probably due to the fact that the fatty acids form rigid complexes with it. The formation of such rigid complexes can be prevented to a certain extent by raising the temperature $> 30°$C and by the introduction of a *cis* double bond. The ineffectiveness of the components with fatty acid chains shorter than 10 C atoms indicates that a minimum number of van der Waal contacts are required for the formation of an enzymatically active lipoprotein complex. The conditions which lipid compounds must fulfill to support enzymatic activity are met by a variety of lipid compounds other than the constituents of the natural phospholipids as shown by the fact that the calcium dependent ATPase activity can be restored by a number of nonionic ionic detergents like Triton X-100 or Brij 58[75, 164, 165]. The detergents, especially Triton X-100, can be used to displace the natural membrane lipids from the membrane and to keep the protein simultaneously in an active conformation. Membrane proteins and lipids have been separated on DEAE cellulose and on Sepharose 4B columns in Triton X-100 containing media which leads to the replacement of endogenous lipids by $\sim$70 molecules of Triton X-100 with the exception of a residue of 3—6 molecules of natural lipids per protein molecule[166]. However, calcium-dependent ATPase activity of these preparations remains stable only when during the isolation procedure calcium concentrations in the millimolar range are present[71, 75]. This stabilizing effect of calcium ions is quite specific; magnesium ions are completely ineffective. In the absence of calcium ions an irreversible molecular arrangement takes place. Under these conditions the protein the ionic strength of the eluant must be raised to elute the protein from the DEAE column. This finding can most easily be explained by the assumption that during the rearrangement of the molecular structure the number of acidic amino acid residues at its surface is increased. Concomitantly, the hydrophobic domain of the surface is reduced giving rise to a release of bound Triton X-100[166]. The ATPase activity of this lipid depleted and structurally modified preparation can, too, be restored with fatty acids provided that the carbon chain is sufficiently long. However, the resulting enzym does not require anymore calcium ions for its activity. In contrast to the calcium-dependent ATPase, the activity of the calcium-independent enzyme does not decline if the length of the saturated fatty acid chain exceeds 12 C-atoms[167]. The formation of rigid complexes is presumably prevented by the detergents present in the assay.

The calcium-independent ATPase of the lipid modified preparations is not only different from the calcium-dependent ATPase but also from the calcium-independent ATPase of native preparations — the basic ATPase — which has a lower nucleotide specificity[126]. The experiments in which the lipid matrix of the sarcoplasmic membranes has been replaced by lipid compounds not present in native membranes reveal a high degree of functional flexibility of the enzyme. On the other hand, a few residual lipids in the protein are sufficient to prevent these changes in the structure of the enzyme and to preserve its calcium sensitivity.

The view that the sarcoplasmic transport protein is a stable and at the same time an adaptable enzyme is supported by results of experiments in which endogenous lipid matrix has been replaced by exogenous well defined phospholipid molecules. Preparations which were partially delipidated by the action of enzymes are not very well suited for studying the reactivation of their ATPase by exogenous phospholipid molecules. Phospholipids are hydrolyzed by these preparations because it is difficult to free them from traces of the applied enzymes[168]. A suitable method for analyzing the interaction of diacyl-phosphatidyl compounds with membrane proteins which avoid lipid removal have been developped by Metcalfe and co-workers[93, 94]. The method involves the direct replacement of endogenous by exogenous lipids dispersed in cholate. If the displacement procedure is repeated two times, the endogenous lipids can be exchanged nearly completely. In other procedures Triton X-100[166] or cholate[160] are applied for nearly complete lipid depletion and subsequent lipid supplementation. All phospholipids tested as substitutes support activity. Optimal activity is achieved with dioleyl-phosphatidylcholine followed by dimyristyl-phosphatidylcholine and dipalmythyl phosphatidylcholine. The activity of the latter, however, becomes discernable only at temperatures above 30°C. Experiments in which the endogenous lipids were exchanged by a mixture of two defined exogenous lipids indicate that the protein does not combine selectively with one of the offered lipids[94]. It must, however, be kept in mind that specific differences might have been obscured by the presence of the detergent. Although the calcium-dependent ATPase of lipid modified preparations resemble the native enzyme, they are not identical. An enhanced sensitivity of the lipid modified enzyme towards stimulating and inhibiting effect produced by monovalent cations reveals some delicate differences[169]. The enhanced ion sensitivity suggests that the ion binding sites have become more flexible as the result of lipid modification or substitution.

While reconstitution of the calcium-dependent ATPase from the lipid deprived enzyme can easily be achieved, attempts to reconstitute simultaneously the abolished accumulation of calcium had no success[55, 70]. Yet, in a number of reports the reconstitution of calcium transport from the enzyme after purification and/or after lipid exchange has been described[160, 170–172]. In these experiments it was attempted to reconstitute vesicles which could retain calcium ions which were transported into the vesicular space by the transport protein across the lipid bilayer. Different lipid protein ratios varying from 0.3 to 20 were applied and different procedures were advocated for the removal of the detergents and for promoting the formation of vesicles. The observed calcium storage must be considered, at least in some cases, to be mainly due to the spontaneous precipitation of calcium oxalate or calcium phosphate because the concentrations of the respective ions in the assay media exceeded the solubility product. The accumulation of a small quantity of calcium ions and the formation of a very few crystals may have started spontaneous precipitation. The most characteristic feature of these reconstituted preparations is their permanently active ATPase which is in contrast to native vesicles whose calcium-dependent ATPase disappears when the pump has removed calcium from the external medium. The permanently high ATPase activity of the reconstituted preparation therefore indicates the inability of this preparation to reduce the calcium concentration below levels which support enzymatic activity. It cannot be decided if this inability is due to a degeneration of the coupling mechanism between ATP splitting and calcium translocation or due to the inability of the reconstituted vesicular membrane to retain the accumulated calcium ions.

5 Elementary Steps of the Reaction Sequence

From the description of the coupling between ion movement and ATP cleavage or synthesis it infers that the transport protein must be able to interact with at least five reactants. In the following, these interactions will be analyzed. To facilitate the analysis, the reaction sequence will be dissected in its elementary steps whenever the analysis does not require the complete transport system. Partial reactions can most favourably be studied with modified preparations or the isolated transport protein. These preparations do not only provide simplicity but enable one to gain inside into the role of the various constituents of the system.

5.1 Substrate Binding

5.1.1 High Affinity Calcium Binding

Calcium ions are bound with an identical high affinity of $5.10^6 M^{-1}$ by the purified ATPase, by the transport protein in the native membranes as well as by partially delipidated, reversibly inactivated membrane preparations[118, 119, 173]. The amount of calcium which is bound with that high affinity corresponds to two sites per transport molecule. The observed affinity is in good agreement with the affinity derived from the dependence on ionized calcium of the activation of calcium uptake and ATP splitting as well as of the inhibition of calcium release and ATP synthesis[18, 112, 174, 175]. Since the latter experiments were performed under conditions which provide a constant internal free calcium concentration by the presence of oxalate or phosphate in the system, the reactions must have been activated or inhibited by the calcium ions in the external solution where their concentration was buffered by calcium EGTA. The two involved calcium binding sites must, therefore, be located at the outer surface of the sarcoplasmic reticulum vesicles. The dependence of the observed activities on the concentration of ionized calcium is described by a Hill coefficient of 2 which indicates that two calcium binding sites must interact cooperatively.

5.1.2 Low Affinity Calcium Binding

Binding sites for calcium with a much lower affinity have been deduced from the activation of calcium uptake and ATP splitting produced by increasing concentrations of oxalate or phosphate when the concentration of calcium in the external medium was kept constant[8, 112]. Oxalate and phosphate in the medium bind and precipitate calcium inside the vesicles and prevent thereby the saturation of internal sites which when occupied, produce inhibition. It is, therefore, most likely that the low affinity binding sites are located at the segment of the transport molecule which faces the internal space of the vesicles. This location is supported by the fact that optimal rates of calcium release and of ATP synthesis are only observed when at internal calcium concentration of approximately $>$ 1 mM the internal low affinity sites are occupied. Calcium binding sites with a similar low affinity were found when the competition between calcium and manganese binding was analyzed and when calcium binding was measured directly[118, 175, 176] (Table 10). The values of the observed affinities also agree with those only deduced from the inhibition of the ATPase by calcium ion[161, 175]. The ratio of the observed affinity constants for calcium binding to external and internal binding sites corresponds quite well to the maximum concentration ratio for calcium ions that can be produced by the pump. Internal as well as external calcium binding sites are susceptible to modifiying effects by other cations. Monovalent cations produce a complex pattern of activation and inhibition depending on their chemical nature as well as on the applied concentrations[161]. When the calcium-dependent ATPase is activated by low concentrations of ionized calcium, all alkaline ions at low concentrations $<$ 0.2 M, with the exception of lithium, produce a marked activation. Higher concentrations of the cations uniformly suppress ATP splitting. When modulation by the ions takes place at the low affinity calcium binding sites, i. e. in the presence of high concentrations of calcium ions, activating effects

Table 10. The binding of divalent ions to native sarcoplasmic reticulum vesicles

Binding class		Number of binding sites per ATPase unit	Association constant $\mathrm{M^{-1}}$			Hill coefficient
			Ca^{2+}	Mg^{2+}	Mn^{2-}	
Independent binding sites	I_1	2	1.8×10^6	2.3×10^2	1.3×10^2	1
	I_2	1	1.0×10^2	6.0×10^3	2.3×10^4	1
	I_3	23	1.3×10^3	8.5×10^2	5.4×10^2	1
Cooperative binding sites	C_1	2	5.7×10^3	5.8×10^2	1.0×10^4	4

prevail for potassium, rubidium and sodium salts even when the concentration reaches 0.8 M. In parallel with the activation of the calcium-dependent ATPase calcium transport is stimulated[157]. Like the monovalent cations, magnesium ions seem to interfere with the high affinity calcium binding sites as follows from the inhibiting effect of magnesium ions at high concentrations on calcium transport and ATP splitting[178].

5.1.3 Magnesium Binding

The analysis of the competition between magnesium and manganese binding to sarcoplasmic reticulum membranes revealed the existence of four different magnesium binding sites. The affinity constants range from $2 \cdot 10^2 \, M^{-1}$ to $6 \cdot 10^3 \, M^{-1}$ [176] (Table 10). There seems to be an additional high affinity binding site from which magnesium cannot be displaced by manganese. Since magnesium ions are presumably involved in several reaction steps, it is difficult to obtain additional information from kinetic analysis of calcium transport or ATP splitting. There is no question that magnesium ions are bound together with ATP as magnesium-ATP for which the enzyme exhibits a high affinity[173]. If calcium transport and calcium-dependent ATP splitting need free ionized magnesium for full activation in addition to ATP-bound magnesium, is an unsolved problem. In contrast, the requirement of ionized magnesium for calcium efflux driven ATP synthesis could unambiguously be determined by using EDTA to establish low free magnesium concentrations in the assays. The reaction is half maximally activated at a concentration of ionized magnesium of 0.1 mM[55]. This high affinity of the protein for magnesium in the reverse reaction contrasts to the low magnesium affinity of the sites involved in the phosphorylation of the transport protein by inorganic phosphate in the absence of calcium ions on both sites of the membrane (comp. 5.2.2.3.[178, 180, 185]).

5.1.4 ATP Binding

ATP as well as ADP are bound by the sarcoplasmic reticulum membrane with appreciable affinity only when magnesium ions are present. It is presumably the respective magnesium salt of the phosphate compounds which combines with the protein. The affinity of magnesium ATP for native sarcoplasmic reticulum membranes as well as for the isolated transport protein has been investigated by measuring directly magnesium ATP binding[173] and also by indirect measurements taking advantage of the ability of the nucleotides to protect surfical thiol groups of the protein against substitution[64, 148, 153]. Delipidation of the membranes which abolishes their enzymatic activity does not interfere with ATP binding[64, 163]. The value of $0.3 \cdot 10^6 \, M^{-1}$ for the affinity constant obtained by appropriate binding studies is 3 to 10 fold higher than those obtained by the indirect procedure. The rather complex dependence of the calcium activated ATPase on the concentration of ATP has been taken as an indication for the involvement of more than one class of ATP binding sites. The calcium dependent ATPase activity of native sarcoplasmic reticulum membranes increases over four decades of substrate concentration starting with 0.1 μM. The activity increase at low concentrations is followed by a less steep activity increment between 0.1 mM and 10 mM[143, 164, 181] (Fig. 13). This kind of activity profile has been taken as an indication that we are dealing with negative cooperativity between two ATP binding sites. The observed variability of the shape of the profile, however, shows that the coupling between the two sites is not constant. Sometimes profiles are obtained which can be described by two independent ATP binding sites. The kind of coupling seems to depend on the nature of the lipid environment of the transport ATPase. After cleaving the phospholipids by phospholipase A_2 or after replacing the lipids by Triton X-100 the activity of the ATPase reaches its maximum already at an ATP concentration of 10 μM indicating that the negative cooperativity is no longer present[55].

5.1.5 ADP Binding

In binding experiments, the affinity of magnesium ADP to native membranes and to the isolated calcium dependent ATPase was found to be considerably lower than that of magnesium ATP[173]. On the other hand, from the inhibition of the calcium-dependent ATPase or the activation of calcium release and ATP synthesis apparent affinities for ADP are obtained that are very similar to those of ATP (Fig. 12). The affinity of ADP for the enzyme apparently depends on its functional state. The affinity of ADP for the membranes under conditions of calcium release depends markedly on the pH of the medium. When the medium pH is reduced from 7.0 to 6.0, the affinity drops by a factor of 10. At pH 7.0 the affinity of the membrane for ADP corresponds to the affinity for ATP to the high affinity binding sites in the forward running mode of the pump. In contrast to the complex dependence of the forward reaction on the concentration of ATP, the dependence of the reverse reaction on ADP seems to follow simple Michaelis-Menten kinetics.

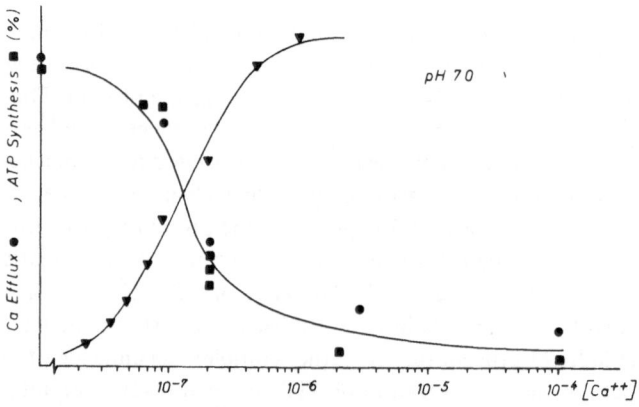

Fig. 12. Activation of calcium uptake and the inhibition of calcium release and ATP synthesis by calcium ions. ▼——— ▼ calcium uptake, ● - - - - ● calcium release, ■ - - - - ■ ATP synthesis

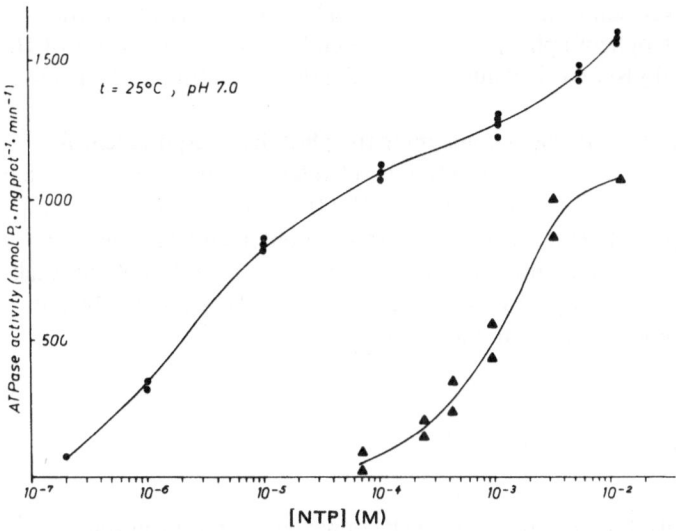

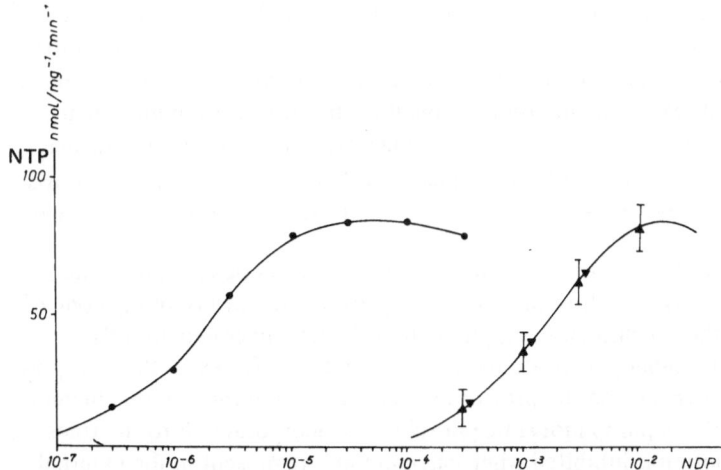

Fig. 13. Dependence on ATP and ITP concentration of calcium dependent phosphate liberation (a) and the corresponding nucleoside triphosphate synthesis on the concentration of ADP, IDP and GDP. 13a: ● ATP, ▲ ITP, 13b: ● ADP, ▲ IDP, ▲ GDP

5.1.6 Binding of Low Affinity Phosphate Donors and Inorganic Phosphate

The affinities of the transport protein for the nucleoside triphosphates ITP and GTP and the corresponding diphosphates deduced from the concentration dependence of either calcium-dependent phosphate liberation or phosphate incorporation during calcium efflux are at least 10 to 400 times lower than the affinity of the adenine nucleotides (Fig. 12). The affinities for acetyl phosphate, carbamyl phosphate, para-nitro-

39

phenyl phosphate are even somewhat lower. Interestingly, the affinity of the transport ATPase for para-nitrophenyl phosphate is increased by more than a factor of 10 in the presence of dimethylsulfoxide while it does not affect the affinity of the nucleotides[143].

There is relatively good agreement concerning the phosphate requirement for calcium release and ADP synthesis. The reaction is activated maximally at a phosphate concentration of approximately 1 mM at 20°C at pH 7.0 when excess magnesium is present[182]. Apparent phosphate affinities have also been obtained from the phosphate concentrations which are required for the inhibition of the hydrolysis of weak substrates like ITP and para-nitrophenyl phosphate. Data for phosphate binding obtained from direct binding studies are not available.

5.2 Phosphoproteins as Reaction Intermediates

In contrast to the hydrolysis and synthesis of ATP connected with proton translocation in mitochondria, chloroplasts and bacterial membranes, the energy linked movement of calcium ions gives rise to the appearance of an acid-stable phosphorylated intermediate in the membranes. A cation specific phosphorylation also occurs in the membranes of the sodium potassium transport system[183]. However, due to the inability to correlate phosphorylation and ion movement in the latter membranes, membrane phosphorylation has been questioned as being a step in the reaction sequence of ion translocation[184, 185]. Solely the sarcoplasmic calcium transport system allows to correlate directly and quantitatively ion translocation with the phosphoryl transfer reactions.

The transporting protein in the sarcoplasmic membrane can be phosphorylated by ATP as well as by inorganic phosphate (cf.[2, 174]). In the forward running mode of the pump, i.e. when the calcium pump accumulates calcium and concomitantly hydrolyzes ATP, the terminal phosphate residue of ATP is transferred to the transport protein. The reaction depends on the presence of calcium ions in the external medium. In the reverse mode of the pump inorganic phosphate is incorporated into the transport protein. This reaction is inhibited when calcium ions are present in the external medium.

5.2.1 Phosphorylation by ATP

A direct involvement of a phosphorylated intermediate in the forward running mode of the pump was first indicated by the finding that during calcium transport a rapid exchange of phosphate between ADP and ATP takes place, and that under various conditions calcium transport and the exchange reaction are concomitantly activated and inhibited[186−188]. Finally, the formation of a phosphoprotein as an intermediate of the exchange reaction could be shown to occur under the conditions of ATP-dependent calcium transport[189−191]. The phosphorylated intermediate was trapped by quenching the reaction with acid. The phosphoryl transfer from ATP is specifically activated by magnesium and calcium ions and proceeds much faster than phosphate

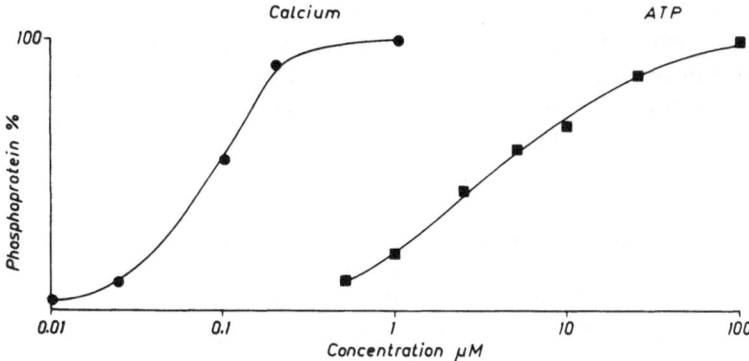

Fig. 14. Dependence of the steady state phosphoprotein level on the concentration of ionized calcium and ATP, respectively. Maximal phosphoprotein levels 3–4 nmol/mg protein. The concentration range of ionized calcium and ATP in which the level of phosphoprotein and the ATPase activity rise are identical (comp. Figs. 12, 13)[177, 191]

liberation during calcium transport. Yet, steady state phosphorylation and the rate of phosphate liberation rise in parallel when the concentrations of ATP and of calcium are increased. For the two substrates this rise occurs in the same concentration range in which the steady state rate of phosphate liberation and calcium uptake increases (Fig. 14). This parallelism strongly indicates that the phosphorylated intermediate is causally involved in calcium transport. In contrast to ATP binding phosphorylation like the calcium-dependent ATPase requires the presence of lipids[85, 160, 161].

The phosphoryl group of the intermediate can enter two different reaction pathways leading to its decomposition. The phosphoryl group can either be transferred to water or to ADP. The hydrolytic pathway leading to the liberation of phosphate must be coupled to calcium translocation as it infers from the fixed coupling between calcium accumulation and phosphate liberation.

Recent findings of Ikemoto[116, 124] have cast some light on the problem how this coupling between phosphoryl transfer and calcium movement might be brought about. Simultaneous measurement of phosphorylation and calcium binding indicates that a second acid-stable phosphorylated intermediate occurs in the reaction chain. This intermediate has a lower affinity to calcium than the first intermediate which is formed when the high affinity calcium binding sites of the transport protein are occupied. Due to the formation of a phosphorylated intermediate with a lower calcium affinity, calcium is released from the protein. It remains to be shown that this transient affinity change is comparable to the affinity difference which is required to explain the concentration of calcium ions by more than a factor of 1000 as it occurs during active calcium accumulation.

5.2.1.1. Hydrolytic Decomposition of Phosphoprotein. The kinetic analysis of phosphate liberation revealed that phosphate is not steadily liberated but that an initial lag period occurs which is followed by a transient burst of phosphate liberation[123, 177, 192]. This burst has been interpreted as resulting from the accumulation of an acid-labile intermediate arising from an acid-stable precursor. The burst is followed by a continuous release of phosphate when the concentration of the intermediate has become constant. While there is no dispute concerning the occurrence of such a burst of phosphate liberation, its size is a matter of conjecture. The assumption that the burst of phosphate liberation results from the formation of an acid-labile intermediate is conclusive only as long as the amount of phosphate liberated during the burst does not exceed the maximum number of phosphate accepting sites in the preparation. This, however, seems to be the case. At higher concentrations of ATP (1 mM) the amount of phosphate liberated during the steady state period is 3 to 5 times higher than the number of available sites from which phosphate could be set free by acid. Therefore, it remains uncertain whether an acid-labelled intermediate (E*P in addition to the acid stable one (E~P) is formed in the following reaction sequence[177].

$$E + ATP \xrightleftharpoons[200 \text{ s}^{-1}]{10^7 \text{ M}^{-1} \text{ s}^{-1}} EATP \xrightleftharpoons[50 \text{ s}^{-1}]{150 \text{ s}^{-1}} E \sim P \xrightleftharpoons[45 \text{ s}^{-1}]{50 \text{ s}^{-1}} \overset{*}{E} \cdot P \xrightleftharpoons{10 \text{ s}^{-1}} \overset{*}{E} + P$$

The stability of the acid-stable phosphoenzyme depends on the presence of both, calcium and magnesium ions in the medium. When the formation of the phosphoenzyme is blocked by removal of calcium ions the intermediate rapidly decays. In contrast, the removal of both ions produces only a partial decay which, however, becomes accelerated on readdition of magnesium ions. These results support the view that magnesium ions are involved in the activation of phosphoprotein decomposition when the pump works in its forward running mode[178].

5.2.1.2. Phosphate Exchange Between ATP and ADP. As mentioned above, the phosphoryl group of the intermediate donated by ATP, can be transferred back to ADP apart from being decomposed by hydrolysis. This phosphoryl transfer to ADP gives rise to a phosphate exchange between ATP and ADP. This exchange can take place between all nucleoside tri- and diphosphates[187]. The ATP-ADP exchange reaction proceeds faster than calcium transport and phosphate liberation. In contrast to the sodium potassium ATPase, an inhibition of the hydrolytic pathway of the calcium transport ATPase by thiol reagents or by the reduction of the magnesium concentration does not result in an enhancement of the exchange reaction. It is rather inhibited in parallel with ATP hydrolysis and calcium transport. This indicates that the corresponding steps in the reaction chain of the two ion transport systems differ as far as their requirements for magnesium and for the presence of functional thiol groups are concerned. Calcium dependent phosphate exchange between ATP and ADP seem to be the most sensitive reaction steps of the complete sequence. Different small modifications of the lipid constituents of the membrane which activate or do not affect the ATPase activity of the preparation severely inhibit the exchange activity and this inhibition is irreversible[169]. Since the first step of the reaction chain, namely the transfer of the terminal phosphate of ATP to the protein, is not or only little depressed, one has to assume that the vulnerability resides in the transfer of the phosphate group to ADP. In contrast to the phosphorylation of the protein, dephosphorylation by ADP proceeds even when calcium and magnesium ions have been removed from the medium by chelation[178]. It has been reasoned that the transfer reaction is catalyzed by calcium ions which enter the reaction center in the membrane from the internal space of the vesicles. In spite of the fact that ADP can be phosphorylated even if the solution contains no magnesium ions, an involvement of firmly bound magnesium is very likely. This is supported by the finding that manganese ions can substitute magnesium ions as far as phosphorylation of the enzyme by ATP, ATP splitting and calcium transport are concerned, but the exchange reaction is not catalyzed by manganese ions[55]. If one does not assume that manganese inhibits the transfer of the phosphoryl group to ADP when bound in addition to magnesium, the displacement of bound magnesium by manganese could be a plausible explanation. The described transfer of the protein bound phosphoryl group to ADP presumably is the terminal step in the reaction sequence when the pump synthesizes ATP during calcium release.

5.2.2. Phosphorylation of the Transport Protein by Inorganic Phosphate

The initial step in the sequence which leads to ATP synthesis during calcium release is the incorporation of inorganic phosphate into the transport protein. It was first demonstrated in experiments with sarcoplasmic reticulum vesicles which were actively loaded with calcium phosphate[193, 194]. The membranes of the calcium loaded vesicles rapidly incorporate inorganic phosphate when the concentration of ionized calcium in the assay is reduced by the addition of EGTA. The involvement of this phosphorylated intermediate in ATP synthesis infers from the finding that on addition of ADP the level of phosphoenzyme drops and simultaneously, calcium is released and ATP is synthesized[193, 194] (Fig. 15). The same observations have been made when the

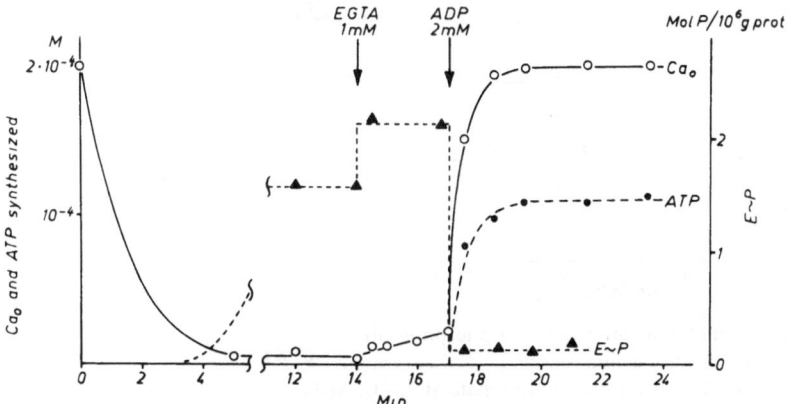

Fig. 15. Incorporation of inorganic phosphate into sarcoplasmic reticulum membranes and the transfer of the phosphoryl group to ADP during calcium release[191]. The incubation medium for calcium loading contained 2.0 mM acetyl phosphate, 7 mM $MgCl_2$, 20 mM potassium phosphate, 0.2 mM $CaCl_2$, 0.5 mg protein/ml. In order to prevent the accumulation of ATP, the system is supplemented by hexokinase and glucose (0.1 M) as final P_i-acceptor. Calcium concentration in the solution and ATP formation, left ordinate: phosphoprotein, right ordinate. When acetyl phosphate is used as energy donor, the protein starts to incorporate inorganic phosphate, even if there is residual calcium in the external solution. On addition of EGTA the phosphoprotein level reaches its maximal value

vesicles were passively loaded with calcium by prolonged incubation in concentrated calcium solutions[157, 194]. These experiments were performed in order to exclude that residual energy donors such as acetyl phosphate used for calcium loading might have contributed to the incorporation of inorganic phosphate. The result that only after calcium loading a substantial membrane phosphorylation was obtained has been taken as evidence that incorporation of phosphate depends on the existence of a calcium gradient across the membranes. The importance of the gradient for the formation of phosphoprotein is most clearly revealed by its spontaneous breakdown when the gradient is abolished. This can most easily be done by the incorporation of the ionophore X 537 A into the membrane[195] (Fig. 16). Furthermore, the phosphoryl group incorporated under the influence of the gradient can readily be transferred to ADP. As to the nature of the phosphorylated intermediate formed by phosphate incorporation, doubts concerning its uniform character arose when Kanazawa and Boyer[180] and Masuda and de Meis[196] found that various sarcoplasmic reticulum preparations of closed and leaky vesicles or even the isolated transport ATPase were phosphorylated in the absence of a calcium gradient, provided that the calcium concentration in the medium was reduced. Yet, under the conditions where in the presence of

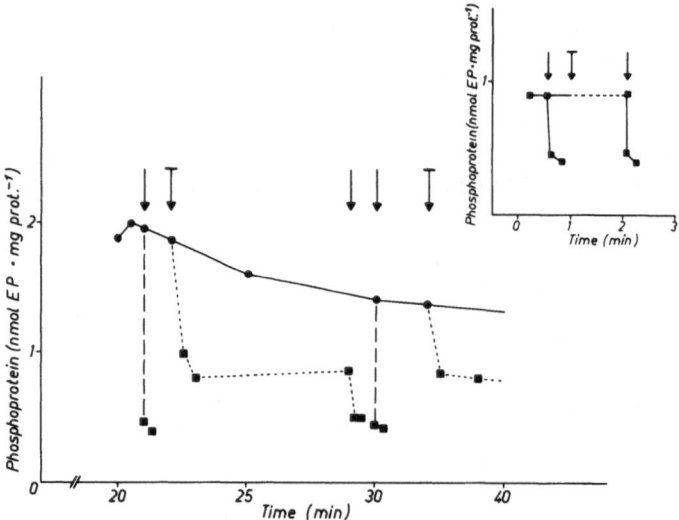

Fig. 16. Decomposition of gradient-dependent phosphoprotein induced by the addition of the calcium ionophore X 537 A and ADP[195]. Native vesicles were loaded with calcium phosphate at 23 °C. After addition of EGTA final concentration 10 mM, the phosphoprotein (●) level rises in a few seconds to a maximum and then slowly declines. On addition of 40 μM ionophore X 537 A (I), phosphoprotein (■ - - - - ■) rapidly decays to the same level which is reached when empty vesicles are phosphorylated in the absence or in the presence of the ionophore (■——■ inset). The final level of the phosphoprotein does not depend on the level reached or present before the addition of the ionophore. Addition of 0.1 mM ADP, effects an even more rapid decay of gradient-dependent phosphoprotein (■ - - - -■). The final phosphoprotein level is identical with the level observed when ADP is added to phosphorylated empty vesicles (■ —— ■ inset)

Table 11. Properties of the phosphorylated intermediates formed when inorganic phosphate is incorporated into the transport protein

Properties	Gradient independent phosphoprotein	Gradient dependent phosphoprotein
Acid stability	+	+
Calcium sensititvity[a]	0.3–0.5 μM	0.3–0.5 μM
Dependence on magnesium ions[b]	10 mM	0.1 mM
Phosphate affinity, Km for $H_2PO_4^-$	10 mM	<0.3 mM
Heat of formation	48.0 KJ mol^{-1}	<12.0 KJ mol^{-1}
Rate of formation (s^{-1}) 10 °C, 5 mM phosphate	~0.5	0.12
Rate of turnover (s^{-1}) 10 °C, 5 mM phosphate	1.0	0.14
Sensitivity to calcium ionophores	None	Yes

[a] Calcium sensitivity denotes the concentration of ionized calcium at which 50% of the phosphoprotein are decomposed at pH 7.0.
[b] Magnesium concentrations which are required for half maximum formation of phosphoprotein.

a gradient 2–3 nmol phosphoenzyme per mg protein were formed by closed vesicles, open membranes incorporate less than 0.4 nmol phosphate per mg protein. To obtain with the latter preparation a high yield of phosphoprotein the pH of the medium must be reduced, its phosphate concentrations and the temperature must be increased and the ionic strength has to be kept low[182, 195]. Furthermore, in the absence of a gradient, phosphorylation requires the presence of high concentrations of magnesium ions (Table 11). As mentioned above, approximately magnesium are sufficient to achieve half maximum phosphorylation in the presence of a gradient. A further important discriminating difference is the fact that the phosphoryl residue incorporated in the absence of a gradient cannot be transferred to ADP. These marked differences strongly indicate that two different classes of phosphoenzyme are formed when calcium is removed from the protein depending on whether calcium ions are present or absent at the internal surface of the vesicles. Calcium removal from the external high affinity calcium binding sites of the vesicular membranes is a common prerequisite for the formation of both phosphoproteins. In the presence as well as in the absence of internal calcium, phosphate incorporation reaches its maximum only when calcium is completely removed from the high affinity binding sites at the external membrane surface or, vice versa, phosphate incorporated at high pCa is displaced when the calcium concentration is elevated again. The relationship between calcium binding and phosphate displacement is pH-dependent[195]. At pH 6.0 calcium binding and phosphate displacement are linearly related and the binding of one calcium ion can be sufficient to release one phosphate residue. In contrast, at pH 7.0 the relationship is quadratic, indicating that a displacement of one phosphate residue requires at least the binding of two calcium ions. Evidently, at pH 7.0 phosphorylation and dephosphorylation of the membrane, respectively, produced either by the removal or the addition of calcium ions reflect the same two to one relationship which exists between calcium uptake and ATP splitting or between calcium efflux and ATP synthesis. The common dependence of the two phosphoprotein species on the calcium concentration in the medium makes it unlikely that we are dealing with completely unrelated phosphorprotein species, a view which might be supported by their contrasting other properties. It is most suggestive to assume that the gradient-independent phosphoprotein is a precursor of the gradient dependent species. This possibility may be explored by the investigation of the energetics and the kinetics of the formation of the two intermediates.

5.2.2.1. Energetics of Phosphoprotein Formation. In the presence of a calcium gradient the magnitude of the available energy for phosphoprotein formation is given by the concentration gradient. When at an external pCa of eight, maximum phosphorylation is reached and simultaneously an internal pCa of three exists, the energy which becomes available when one mol of ions moves downhill is

$$\Delta G = R \cdot T \cdot \ln \frac{Ca_0}{Ca_i} = -28 \text{ kJ/mol}.$$

Since phosphorylation and calcium movement are coupled by a one to two relationship, -56 kJ/mol are provided by the gradient for the formation of one mol of phosphoprotein.

In the absence of a calcium gradient the energy required for the incorporation of phosphate must be provided by the removal of calcium from the protein alone. Since the protein and the calcium chelator EGTA have approximately the same affinity for calcium, the driving force for phosphate binding originates from the reduction of the concentration of ionized calcium alone.
The reaction
$ECa + EGTA \rightleftharpoons E + CaEGTA$

can yield a free energy increment of

$$\Delta G = R \cdot T \cdot \ln \frac{(ECa)_i}{(EGTA)_i}$$

which can drive phosphate incorporation.
The initial concentrations of $(ECa)_i$ and $(EGTA)_i$ are usually $\sim 10 \ \mu M$ and 5 mM, respectively. Under these conditions 12 kJ/mol become available when one mol of calcium ions is removed from the protein. This is scarcely sufficient for the formation of a sizeable fraction of a phosphoenzyme with an acyl phosphate bond.

Even if the incorporation of one phosphate residue is linked to the removal of two calcium ions the energy requirement remains marginal as compared to the energy amount available when at the same pCa a calcium ion concentration of 1 mM exists inside the vesicles.

5.2.2.2. Kinetics of Phosphoprotein Formation. The kinetic comparison of phosphate incorporation in the presence or in the absence of a calcium gradient is made difficult by the high velocity with which the reaction proceeds at temperatures between 20° and 30 °C. When the rate of phosphoprotein formation is reduced by lowering the temperature to $10°C$, phosphoprotein formation in the presence of a calcium gradient can be measured reliably. Yet, it is very difficult to measure gradient-independent phosphoprotein formation. Because of the large positive enthalpy change of the reaction the steady state level of phosphoprotein becomes very small at $10°C$[179, 195, 197]. This is in contrast to the level of gradient-dependent phosphoprotein which remains nearly constant since its heat of formation is much smaller[195].

5.2.2.3. Gradient-Independent Phosphoprotein Formation. At given concentration of phosphate and magnesium the rate of phosphate incorporation depends on whether phosphorylation is started by the addition of EGTA to the assay containing magnesium and phosphate or by the addition of phosphate or magnesium to the protein made calcium free in advance by preincubating it in EGTA containing media. The protein is phosphorylated considerably faster when it has been made calcium free prior to the addition of magnesium or phosphate than when phosphorylation is started together with calcium removal[197]. This finding indicates that on addition of EGTA the phosphate accepting configuration of the protein is formed slowly. This may either be due to a slow dissociation of calcium from the protein or due to a slow change in the protein configuration after the protein has become rapidly calcium free. This process apparently depends crucially on a very specific cooperation between membrane proteins and lipids. Neither lipid depleted preparations nor preparations having a reconstituted lipid matrix incorporate significant amounts of phosphate[55, 197].

The calcium free protein exhibits saturation kinetics for both phosphate and magnesium ions (Fig. 17a). This behaviour excludes a one step phosphorylating mechanism. The simplest possibility is a two step reaction sequence.

$$MgE + P \; \underset{k_{-1}}{\overset{k_1}{\rightleftharpoons}} \; MgE * P \; \underset{k_{-2}}{\overset{k_2}{\rightleftharpoons}} \; MgE - P.$$

$Mg * P$ represents an acid labile intermediary complex. If it is assumed that $k_{-1} \gg k_2$, $k_1/(k_1 + k_2) \sim k_1/k_{-1} = K$, the following relationship for the rate of MgE-P formation is obtained.

$$\frac{1}{V} = \frac{1}{k_2} + \left(\frac{1}{K \cdot k_2} \right) \cdot \frac{1}{P}$$

and the dependence of the steady state level of $MgE - P$ or the phosphate concentration is given by

$$\frac{MgE_0}{MgE - P} = (k_2 + k_{-2})/k_2 + k_{-2}/k_2 \cdot P \cdot K$$

Fig. 17b, c demonstrate the obtained linear relations between the phosphate concentration and the rate as well as the level of phosphoprotein in double reciprocal plots.

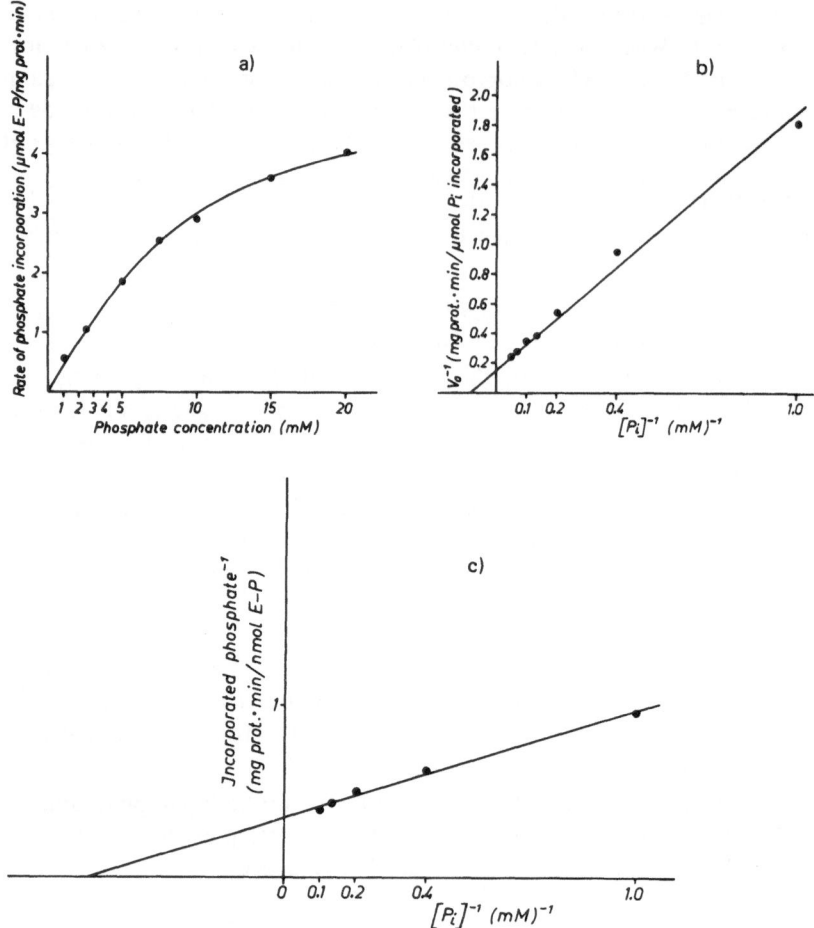

Fig. 17. Dependence of the initial rate of phosphoprotein formation and of the steady state level of phosphoprotein on the phosphate concentration.
a) Phosphate concentration and phosphoprotein formation are linearly plotted.
b) Results of a) are plotted double reciprocally.
c) Double reciprocal plot of phosphoprotein level versus phosphate concentration. Conditions: 10 mM $MgCl_2$, 5 mM EGTA, 10 mM phosphate, pH 6.0, T = 30 °C. Phosphorylation was started by the addition of $MgCl_2$ to the otherwise complete assay

Measurements of the steady state phosphoprotein level at different temperatures revealed that phosphoprotein formation is accompanied by a large and constant enthalpy change of 48 kJ/mol. In contrast, the likewise quite high activation energy of phosphoprotein formation exhibits a pronounced break between 20°C and 30°C. A break in the Arrhenius plot of the calcium-dependent ATPase has been observed in the same temperature range and has been interpreted as transitions between two activity states of the enzyme. Apparently, the phosphorylation of the calcium free protein by inorganic phosphate exhibits a similar kind of activity transition as observed for the calcium-dependent interaction of the transport protein with ATP[131]. A similar transition phenomenon complicates the time course of phosphoprotein formation between 20°C and 30°C While at 20°C and low phosphate concentrations phospho-protein formation occurs monophasically, it becomes biphasic at high phosphate concentrations at 20 °C as well as at 30 °C. Approximately half of the protein is phosphorylated with a half time of 10–20 ms at 30 °C, while phosphorylation of the other half needs 10 times longer. The same biphasic behaviour is found when the rate of turnover of the phosphate residue is measured by addition of tracer amounts of radioactive phosphate after phosphorylation by inactive phosphate has reached steady state level. According to the assumed mechanism, the half time of turnover reflects the spontaneous decay rate of the acid stable intermediate. The biphasic decay indicates the presence of two different intermediates. The mean half time at 30°C is 30 ms and agrees quite well with a value observed by Boyer et al.[198]. The view of the existence of two classes of phosphoprotein is further supported by the biphasic time course which is observed when the once formed phosphoprotein is decomposed by the addition of high concentration of calcium ions (Fig. 18). Both phases of phosphate liberation proceed 10 times slower than the rates of spontaneous release observed in the turnover experiments. One is, therefore, tempted to assume that the addition of calcium produces a transient stabilization of the phosphoprotein species before it causes phosphate release[197].

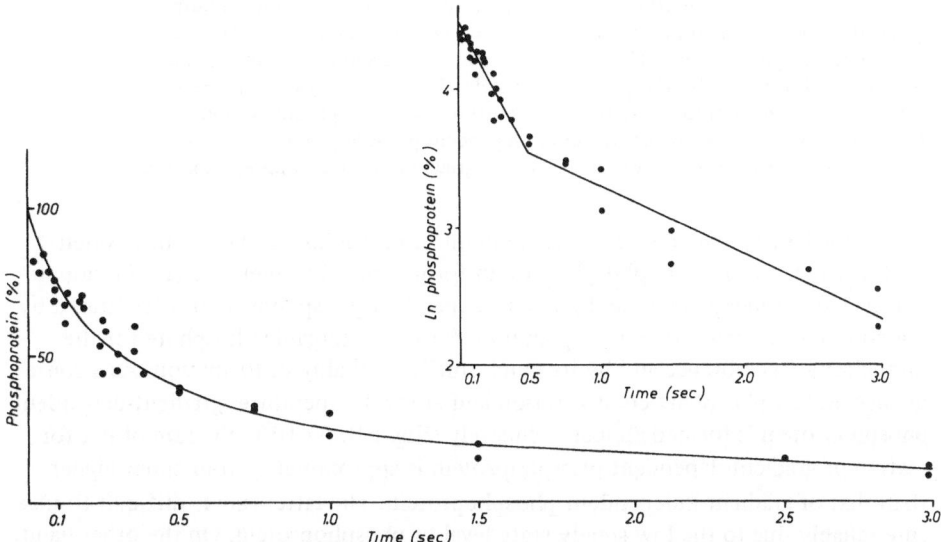

Fig. 18. Time course of the decay of phosphoprotein induced by the addition of calcium chloride[197] 11 mM $CaCl_2$ were added to the phosphorylation medium containing 5 mM EGTA, 5 mM inorganic phosphate, 5 mM glycerophosphate, pH 6.0, T = 30 °C. Note the slow biphasic decay of the phosphoprotein

5.2.2.4. Gradient-Dependent Phosphoprotein Formation. Properties of gradient-dependent phosphoprotein are more complicate to study than those of the gradient-independent phosphoprotein species. In order to obtain a maximum yield of gradient-dependent phosphoprotein of $\sim$ 3 nmol/mg, it is important to accomplish that as

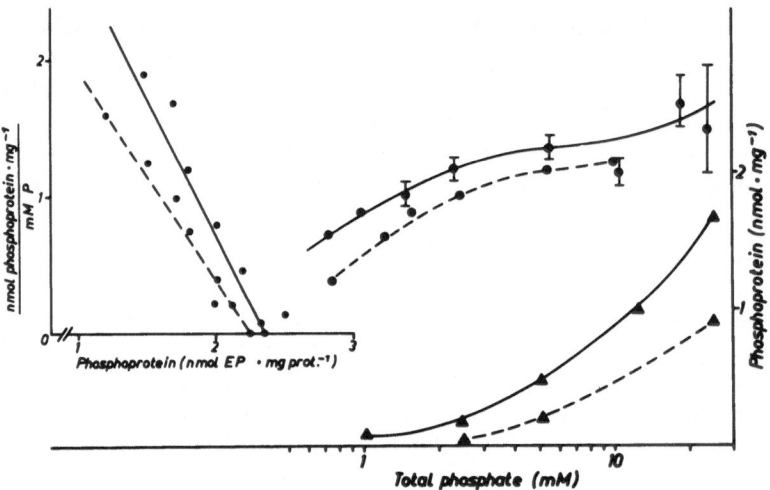

Fig. 19. Dependence of phosphoprotein formation in the presence and the absence of a calcium gradient on the concentration of phosphate [195]. Native vesicles were loaded with calcium phosphate at 23 °C in the presence of different phosphate concentrations. Formation of phosphoprotein was started by the addition of EGTA at a final concentration of 10 mM. The reaction was quenched after one minute when gradient-dependent phosphoprotein has reached an optimal value. The formation of phosphoprotein at 10 °C was started by EGTA addition after the assay was cooled. During the cooling period the vesicles do not lose any calcium, even if the assay is kept cool for more than 5'. Gradient-dependent phosphoprotein ● —— ● 23 °C; ● - - - - ● 10 °C; gradient-independent phosphoprotein ▲——▲ 23 °C; △ - - - - △ 10 °C. Inset: scatchard plot for gradient-dependent phosphoprotein

many vesicles as possible in the assay contain some calcium at the moment when EGTA is added to initiate phosphoprotein formation. The kinetic discrimination between gradient-independent and gradient-dependent phosphoprotein is facilitated by the considerable increase of the apparent affinity of inorganic phosphate for the sarcoplasmic membranes and by its much smaller enthalpy of formation. As a consequence, at low phosphate concentration and at low temperatures gradient-dependent phosphoprotein is formed almost exclusively (Fig. 19). At 10°C the rate of net formation of gradient-dependent phosphoprotein is approximately four times higher than that of gradient-independent phosphoprotein. The latter rate is difficult to measure reliably due to the low steady state level of phosphoprotein. On the other hand, measurements of the rate of phosphate exchange revealed that the dissociation of the gradient-dependent phosphoprotein proceeds definitely slower than that of the gradient-dependent species.

A more conclusive kinetic analysis is not only made difficult because of the temperature-dependent occurrence of different phosphoprotein species, but also by the fact that the spontaneous decay of gradient-independent as well as of gradient-dependent phosphoprotein depends on the concentration of phosphate. This phosphate dependence is not taken into account by the tentative reaction scheme.

Since the properties of the two phosphorylated intermediates make it difficult to find experimental conditions under which both species exist simultaneously in sufficient quantities, it is not possible to analyze the interrelationship of the two intermediates at the moment. Therefore, it cannot be decided whether gradient-independent or gradient-dependent phosphoproteins are intermediates of the same reaction chain if they are products of a reaction proceeding in parallel or if we are dealing with a branched reaction chain. If we assume that the gradient-dependent and the gradient-independent phosphoprotein belong to the same reaction chain, gradient-independent phosphoprotein would be transformed to gradient-dependent phosphoprotein when calcium is present at the internal surface of the vesicles. For two separate reactions occurring in parallel, starting from a common unphosphorylated protein precursor, we must postulate that two states of the protein exist depending on internal calcium which differ in their affinity for phosphate. In a branched reaction pathway a common intermediate might be formed whose transformation depends on the presence or the absence of internal calcium ions.

In any case, gradient-dependent phosphoprotein can only be formed when internal low affinity calcium binding sites are occupied and external high affinity calcium binding sites are not occupied. The question arises whether the different saturation of internal and external binding sites by calcium ions is sufficient for the formation of gradient-dependent phosphoprotein or whether it is necessary that calcium ions are translocated from the low to the high affinity sites across the membranes. The undisputed close coupling between the transition of the calcium across the membranes and continuous ATP formation can be explained by both mechanisms. If an asymmetric calcium distribution is sufficient for gradient-dependent phosphoprotein formation, one could imagine that calcium additon to small membrane fragments of purified ATPase preparations can transiently give rise to such a calcium distribution provided that calcium ions are bound more rapidly to internal than to external binding sites. The fact that a great fraction of the gradient-independent phosphoprotein decays on addition of calcium slower than phosphoprotein formation proceeds during calcium removal is a most striking property of this compound. Since the calcium-dependent decay of the phosphoprotein also proceeds much slower than its spontaneous decay, one is tempted to assume that the addition of calcium produces a transient stabilization of phosphoprotein. If this long living intermediate would have become ATP-sensitive it can possibly give rise to ATP formation provided that ADP accepts the phosphoryl group sufficiently fast. This kinetic property of the phosphoprotein can be considered as a basis for ATP formation by nonvesicular preparations produced by the simultaneous addition of high concentrations of calcium ADP to the phosphorylated enzyme [199, 200].

Note Added in Proof

The alternative discussed on p. 47 could be decided in the meantime: The rate of phosphorylation of the protein by inorganic phosphate is limited by the slow dissociation of the calcium protein complex. Vice versa, when phosphorylation of the protein by ATP is started by the addition of calcium, the slow formation of the calcium protein complex limits the rate of phosphorylation [201].

6 References

1. Hasselbach, W.: Progr. Biophys. Mol. Biol. *14*, 167–222 (1964)
2. Hasselbach, W., in: Enzymes X. New York: Acad. Press 1974, pp. 431–467
3. Inesi, G.: Ann. Rev. Biophys. Bioeng. *1*, 191–210 (1972)
4. MacLennan, D. H., Holland, P. C.: Ann. Rev. Biophys. Bioeng. *4*, 377–404 (1975)
5. Krogh, A.: Z. vergl. Physiol. *25*, 335–356 (1938)
6. Harris. J. E.: J. Biol. Chem. *141*, 579–595 (1941)
7. Netter, H.: Theoretische Biochemie. Berlin, Göttingen, Heidelberg: Springer 1959
8. Bernard, C., in: Founders of Experimental Physiology. Boylan, J. W. (Ed.), München: J. F. Lehmann 1971, pp. 185–196
9. Harris, E. J., in: Transport and Accumulation in Biological Systems. London: Butterworths, 1972
10. Glynn, I. M., Karlish S. J. D.: Ann. Rev. Physiol. *37*, 13–55 (1975)
11. Katz B., in: Nerv, muscle and synapse. London: McGraw-Hill 1966
12. Simons, J. B.: J. Physiol. (London) *206*, 227–244 (1976)
13. Portzehl, H., Caldwell, P. C., Rüegg, J. C.: Biochim. Biophys. Acta *79*, 581–591 (1964)
14. Keynes, R. D., Rogas, E., Taylor, R. E., Vergara, J.: J. Physiol. (London) *229*, 409–455 (1973)
15. Baker, P F., Blaustein, M. P., Hodgkin, A. L., Steinhardt, R. A.: J. Physiol. (London) *200*, 431–458 (1969)
16. Ebashi, S.: Ann. Rev. Physiol. *38*, 293–313 (1976)
17. Fuchs, F.: Ann. Rev. Physiol. *36*, 461–502 (1974)
18. Hasselbach, W., in: Molecular basis of motility. Berlin, Heidelberg, New York: Springer 1976 pp. 81–92
19. Reuter, H.: Progr. Biophys. *26*, 1–43 (1972)
20. Schatzmann, H. J.: Experientia *22*, 364–368 (1966)
21. Caldwell, P C., Hodgkin, A. L., Keynes, R. D., Shari, T. I.: J. Physiol. (London) *152*, 561–590 (1960)
22. Dunham, E. T., Glynn, I. M.: J. Physiol. (London) *156*, 274–293 (1961)
23. Hoffman. J. F.: Fed. Proc. *19*, 127 (1960)
24. Whittam, R.: J. Physiol. *140*, 479–497 (1958)
25. Post, R. L.: Fed. Proc. *18*, 480 (1959)
26. Hasselbach, W., Makinose, M.: Biochem. Z. *333*, 518–528 (1961)
27. Skou, J. C.: Progr. in Biophys. *14*, 133–166 (1964)
28. Goldin, S. M., Tong, S. W.: J. Biol. Chem. *249*, 5907–5915 (1974)
29. Hokin, L. E., in: Biochemistry of membrane transport. Berlin, Heidelberg, New York: Springer 1977, pp. 374–388
30. Schatzmann, H. J.: Helv. Physiol. et Pharmacol. *11*, 346–354 (1953)
31. Mitchell, P.: Nature *191*, 144–148 (1961)
32. Jagendorf, A. T., Uribe, E.: Proc. Nat. Acad. Sci., USA *55*, 170–177 (1966)
33. Boyer, P. D., Cross, R. L., Momsen, W.: Proc. Nat. Acad. Sci. *70*, 2837–2839 (1973)
34. Huxley, H. E., in: Molecular basis of motility. Heilmeyer, I. M. G., Rüegg, J. C., Wieland, Th., (Ed.) Berlin, Heidelberg, New York: Springer 1976, pp. 10–24
35. Dancker, P.: Fortschr. Zoologie *25*, 1–71 (1978)
36. Bennett, H. S., in: Structure and Function of Muscle I. Bourne (Ed.) New York: Acad. Press 1960, pp. 137–184
37. Verratti, E., in: Memorie 1st Lomb. Cl. di sc. et nat. *19*, 87–133 (1902)
38. Bennett, H. S., Porter, K. R.: Am. J. Anat. *93*, 61–105 (1953)
39. Wangh, R. A., Spray, T. L., Sommer, J. R.: J. Cell. Biol. *59*, 254–260 (1973)
40. Kielley, W. W., Meyerhof, O.: J. Biol. Chem. *242*, 4637 (1948)
41. Ulbrecht, M.: Biochim. Biophys. Acta *57*, 455–474 (1962)
42. Marsh, B. B.: Biochim. Biophys. Acta *9*, 247–260 (1952)
43. Portzehl, H.: Biochim. Biophys. Acta *26*, 373–374 (1957)
44. Ebashi, F.: Arch. Biochem. Biophys. *76*, 410–423 (1958)

45. Nagai, T., Makinose, M., Hasselbach, W.: Biochim. Biophys. Acta *43*, 223–238 (1960)
46. Agostini, B., Hasselbach, W.: Histochemie *28*, 55–67 (1971)
47. Hasselbach, W.: Fed. Proc. *23* 909–912 (1964)
48. Costantin. L. L., Franzini-Armstrong, C., Podolski, R. J.: Science *147*, 158–160 (1964)
49. Hasselbach, W., Makinose, M.: Biochem. Z. *339*, 94–111 (1963)
50. Heuson-Stiennon, J., Wanson, J., Drochmans, P.: J. Cell Biol. *55*, 471–488 (1972)
51. Weber, A., Herz, R.: J. Gen. Physiol. *52*, 750–759 (1968)
52. Fairhurst, A. S., Hasselbach, W.: Europ. J. Biochem. *13*, 504–509 (1970)
53. Martonosi, A.: J. Biol. Chem. *243*, 71–81 (1968)
54. De Meis, L., Hasselbach, W.: J. Biol. Chem. *246*, 4759–4763 (1971)
55. Hasselbach, W.: unpublished
56. Hasselbach, W., Elfvin, L. G.: J. Ultrastruct. Res. *17*, 598–622 (1967)
57. Martonosi, A.: Biochim. Biophys. Acta *150*, 694–704 (1968)
58. Deamer, D. W., Baskin, R. J.: J. Cell Biol. *42*, 296–307 (1969)
59. Packer, L., Mehard, C. W., Meissner, G., Zahler, W. L., Fleischer, S.: Biochim. Biophys. Acta *363*, 159–181 (1974)
60. Malau, N. T., Sabbadini, R., Scales, D., Inesi, G.: FEBS Lett. *60*, 122–125 (1975)
61. Jilka, R. L., Martonosi, A., Tillach. T. W.: J. Biol. Chem. *250*, 7511–7524 (1975)
62. Martonosi, A., Nakamura, H., Jilka, R. L., Vanderkooi, J. H.: Proc. in Life Science, FEBS Symp. No. 42. Berlin, Heidelberg, New York: Springer 1977, pp. 401–415
63. Hasselbach, W., Seraydarian, K.: Biochem. Z. *345*, 159–172 (1966)
64. Hasselbach, W., in: Mol. bioenergetics and macromolecular biochemistry. Berlin, Heidelberg, New York: Springer 1972, pp. 149–171
65. MacFarland, B. H., Inesi, G.: Arch. Biochem. Biophys. *145*, 456–464 (1971)
66. Martonosi, A., Halpin, R. A.: Arch. Biochem. Biophys. *144*, 66–77 (1971)
67. Meissner, G., Fleischer, S.: Biochim. Biophys. Acta *241*, 356–378 (1971)
68. Louis, C., Shorter, E. M.: Arch. Biochem. Biophys. *153*, 641–655 (1972)
69. MacLennan, D. H.: J. Biol. Chem. *245*, 4508–4518 (1970)
70. Deamer, D. W.: J. Biol. Chem. *248*, 5477–5485 (1973)
71. Ikemoto, N., Bhatnagar, G. M., Gergely, J.: Biochem. Biophys. Res. Communic. *44*, 1510–1517 (1971)
72. Hardwicke, P. M., Green, N. M.: Europ. J. Biochem. *42*, 183–193 (1974)
73. Le Maire, M., Møller, J. V., Tanford, C.: Biochemistry *15*, 2336–2342 (1976)
74. Le Maire, M., Jørgensen, K. J., Røigaard, H., Møller, J. V.: Biochemistry *15*, 5805–5812 (1976)
75. Migala, A., Hasselbach, W.: FEBS Lett. *26*, 20–24 (1972)
76. MacLennan, D. H., Wong, P. T.: Proc Nat. Acad. Sci., USA *68*, 1231–1235 (1971)
77. MacLennan D. H., Yip, C. C., Iles, G. H., Seeman, P.: Cold Spring Harbor Symp. Quant. Biol. *37*, 469–478 (1972)
78. Ostwald, T. J., MacLennan, D. H.: J. Biol. Chem. *249*, 974–979 (1974)
79. Ikemoto, N., Bhatnagar, G. M., Nagy, B., Gergely, J.: J. Biol. Chem. *247*, 7835–7837 (1972)
80. Martonosi, A.: Fed. Proc. *23*, 913–921 (1964)
81. Balzer, H., Makinose, M., Hasselbach, W.: Naunyn Schmiedeberg's Arch. Pharmakol. *260*, 444–455 (1968)
82. Waku, K., Nakazawa, Y.: J. Biochem. (Tokyo) *56*, 95–96 (1964)
83. Owens, K., Ruth, R. C., Weglicki, W. B.: Biochim. Biophys. Acta *288*, 479–481 (1972)
84. Drabikowski, W., Sarzala, M. G., Wroniszewska, A., Lagwinska, E., Drzewiecka, B.: Biochim. Biophys. Acta *274*, 158–170 (1972)
85. Fiehn, W., Hasselbach, W.: Europ. J. Biochem. *13*, 510–518 (1970)
86. Seiler, D., Hasselbach, W.: Europ. J. Biochem. *21*, 385–387 (1971)
87. Boland, R., Chyn, T., Roufa, D., Reyes, E., Martonosi, A.: Biochim. Biophys. Acta *489*, 349–359 (1977)
88. Madeira, V. M. C., Antunes-Madeira, M. C.: Can. J. Biochem. *54*, 516–520 (1976)
89. Scandella, C. J., Devaux, P., McConnell, H. M.: Proc. Nat. Acad. Sci., USA *69*, 2056–2060 (1972)

W. Hasselbach

90. Migala, A., Agostini, B., Hasselbach, W.: Z. Naturforsch. *28c*, 178−182 (1973)
91. Vale, M. G. P.: Biochim. Biophys. Acta *471*, 39−48 (1977)
92. Hidalgo, C., Ikemoto, N.: J. Biol. Chem. *252*, 8446−8454 (1977)
93. Warren, G. B., Toon, P. A., Birdsall, N. J., Lee, A. G., Metcalfe, J. B.: Biochemistry *13*, 5501−5507 (1974)
94. Warren, G. B., Houslay, M. D., Metcalfe, J. C., Birdsall, N. J. M.: Nature *255*, 684−687 (1975)
95. Davis, D. G., Inesi, G., Gulik-Krzywicki: Biochemistry *15*, 1271−1276 (1976)
96. Hasselbach, W., Migala, A., Agostini, B.: Z. Naturforsch. *30c*, 600−607 (1975)
97. Hasselbach, W., Migala, A.: Z. Naturforsch. *30c*, 681−683 (1975)
98. Dupont, Y., Harrison, S. C., Hasselbach, W.: Nature *244*, 555−558 (1973)
99. Herbette, L., Marquardt, J., Scarpa, A., Blasie, J. K.: Biophys. *20*, 245−272 (1977)
100. Stromer, M., Hasselbach, W.: Z. Naturforsch. *31c*, 703−707 (1976)
101. Thorley-Lawson, D. M., Green, N. M.: Europ. J. Biochem. *40*, 403−413 (1973)
102. Stewart, P. S., MacLennan, D. H.: J. Biol. Chem. *249*, 985−993 (1974)
103. Louis, C. F., Buonaffina, R., Binks, B.: Arch. Biochem. Biophys. *161*, 83−92 (1974)
104. Inesi, G., Scales, D.: Biochemistry *13*, 3298−3306 (1974)
105. Martonosi, A., Fortier, F.: Biochem. Biophys. Res. Comm. *60*, 382−389 (1974)
106. Stewart, P. S., MacLennan, D. H., Shamoo, A. E.: J. Biol. Chem. *251*, 712−719 (1976)
107. Meissner, G.: Biochim. Biophys. Acta *389*, 51−58 (1975)
108. Weber, A., Herz, R., Reiss, J.: Biochem. Z. *345*, 329−369 (1966)
109. Carvalho, A. P.: J. Cell. Physiol. *67*, 73−84 (1966)
110. Carvalho, A. P., Leo, B.: J. Gen. Physiol. *50*, 1327−1352 (1967)
111. Mermier, P., Hasselbach, W.: Z. Naturforsch. *30c*, 593−599 (1975)
112. Makinose, M., Hasselbach, W.: Biochem. Z. *343*, 360−382 (1965)
113. Mermier, P., Hasselbach, W.: Europ. J. Biochem. *69*, 79−86 (1976)
114. Martonosi, A., Feretos, R.: J. Biol. Chem. *239*, 648−658 (1964)
115. Scarpa, A., Baldassare, J., Inesi, G.: J. Gen. Physiol. *60*, 735−749 (1972)
116. Ikemoto, N.: J. Biol. Chem. *251*, 7275−7277 (1976)
117. Duggan, P. F., Martonosi, A.: J. Gen. Physiol. *56*, 147−167 (1971)
118. Fiehn, W., Migala, A.: Europ. J. Biochem. *20*, 245−248 (1970)
119. Chevallier, J., Butow, R. A.: Biochemistry *10*, 2733−2737 (1971)
120. Hasselbach, W., Fiehn, W., Makinose, M., Migala, A., in: Mol. basis of membrane function. New York: Prentice Hall 1969, pp. 299−316
121. Beil, F. U., von Chak, D., Hasselbach, W., Weber, H. H.: Z. Naturforsch. *32c*, 281−287 (1977)
122. McComas, Jr., W. H. Riemann, W.: J. Am. Chem. Soc. *64*, 2946−2947 (1942)
123. Kurzmack, M., Inesi, G.: FEBS Lett. *74*, 35−37 (1977)
124. Ikemoto. N.: J. Biol. Chem. *250* 7219−7224 (1975)
125. Barlogie, B.: Med. Thesis Univ. Heidelberg 1972
126. Makinose, M., The, R.: Biochem. Z. *343*, 383−393 (1965)
127. Inesi, G : Science *171*, 901−903 (1971)
128. De Meis, L.: J. Biol. Chem. *244*, 3733−3739 (1969)
129. Pucell, A., Martonosi, A.: J. Biol. Chem. *246*, 3389−3397 (1977)
130. De Meis, L., Masuda, H.: Biochemistry *13*, 2057−2062 (1974)
131. Inesi, G., Cohen, J. A., Coan, C. R.: Biochemistry *15*, 5293−5298 (1976)
132. Barlogie, B., Hasselbach, W., Makinose, M.: FEBS Lett. *12*, 267−268 (1971)
133. Makinose, M., Hasselbach, W.: FEBS Lett. *12*, 271−272 (1971)
134. Makinose, M.: FEBS Lett. *12*, 269−270 (1971)
135. Deamer, D. W., Baskin, R. J.: Arch. Biochem. Biophys. *153*, 47−54 (1972)
136. Panet, R., Selinger, Z.: Biochim. Biophys. Acta *255*, 34−42 (1972)
137. Yamada, S., Sumida, M., Tonomura, Y.: J. Biochem. (Tokyo) *72*, 1537−1548 (1972)
138. Hasselbach, W., Suko, J.: Biochem. Soc. Spec. Publ. Bronk, J. R. (Ed.) 1974, pp. 159−173
139. Temple, D., Hasselbach, W., Makinose, M.: Naunyn Schmiedeberg's Arch. Pharmacol. *282*, 187−194 (1974)
140. Suko. J., Winkler, F., Scharinger, B., Hellmann, G.: Biochim. Biophys. Acta *443*, 517−586 (1976)

141. The, R., Hasselbach, W.: Europ. J. Biochem. *53*, 105–113 (1975)
142. Schulz, G., Elzinga, M., Marx, I., Schirmer, R. H.: Nature (London) *250*, 120–123 (1974)
143. The, R., Hasselbach, W.: Europ. J. Biochem. *74*, 611–621 (1977)
144. Hasselbach, W., Makinose, M., Migala, A.: FEBS Lett. *20*, 311 (1972)
145. Inesi, G., Scarpa, A.: Biochemistry *11*, 356–359 (1972)
146. Hasselbach, W., Seraydarian, K.: Biochem. Z. *345*, 159–172 (1966)
147. Swoboda, G., Hasselbach, W.: Hoppe-Seyler's Z. Physiol. Chem. *354*, 1611–1618 (1973)
148. Panet, R., Selinger, Z.: Europ. J. Biochem. *14*, 440–444 (1970)
149. Green, N. M.: FEBS Meeting, Copenhagen; Abstr. (1977)
150. Dupont, Y., Hasselbach, W.: Nature *246*, 41–43 (1973)
151. Murphy, A. J.: Biochemistry *15*, 4492–4492 (1976)
152. Andersen, J. P., Møller, J. V.: Biochim. Biophys. Acta *485*, 188–202 (1977)
153. Eckert, K., Grosse, R., Levitsky, D. O., Kuzmin, A. V., Smirnov, V. N., Repke, K. R. H.: Acta biol. med. germ. *36*, 1–10 (1977)
154. Ikemoto, N.: personal communication (1977)
155. Vanderkooi, J. M., Martonosi, A.: Arch. Biochem. Biophys. *147*, 632–646 (1971)
156. Inesi, G., Goodman, J. J., Watanabe, S.: J. Biol. Chem. *242*, 4637–4643 (1967)
157. Duggan, P. F.: J. Biol. Chem. *252*, 1620–1627 (1977)
158. Wulf, J., Pohl, W. G.: Biochim. Biophys. Acta *465*, 471–485 (1977)
159. Hesketh, T. R., Smith, G. A., Houslay, M. D., McGill, K. A. Birdsall, N. J. M., Metcalfe, J. C., Warren, G. B.: Biochemistry *15*, 4145–4151 (1976)
160. Knowles, A. F., Eytan, E., Racker, E.: J. Biol. Chem. *251*, 5161–5165 (1976)
161. The, R., Hasselbach, W.: Europ. J. Biochem. *39*, 63–68 (1973)
162. Stoffel, W., Zierenberg, O., Scheffers, H.: Hoppe-Seyler's Z. Physiol. Chem. *358*, 865–882 (1977)
163. The, R.: Proc. Internat. Union of Physiol. Sciences Vol. XIII, 749 (1977)
164. The, R., Hasselbach, W.: Europ. J. Biochem. *30*, 318–324 (1972)
165. Dean, W. L., Tanford, C.: J. Biol. Chem. *252*, 3551–3553 (1977)
166. Green, N. M., in: Calcium transport in contraction and secretion. Carafoli, E. (Ed. *et al.*) Amsterdam: North Holland Publishing Company 1975, pp. 339–348
167. Walter, H., Hasselbach, W.: Europ. J. Biochem. *36*, 110–119 (1973)
168. Roelofsen, B., Schatzmann, H. J.: Biochim. Biophys. Acta *464*, 17–36 (1977)
169. Hasselbach, W., Suko, J., Stromer, M. H., The, R.: Ann. New York Acad. Sci. *264*, 335–349 (1975)
170. Warren, G. B., Toon, P. A., Birdsall, N. J., Lee, A. G., Metcalfe, J. C.: Proc. Nat. Acad. Sci., USA *71*, 622–626 (1974)
171. Meissner, G., Fleischer, S.: J. Biol. Chem. *249*, 302–309 (1974)
172. Knowles, A. F., Racker, E.: J. Biol. Chem. *250*, 3538–3544 (1975)
173. Meissner, G.: Biochim. Biophys. Acta *298*, 906–926 (1973)
174. Hasselbach, W.: Biochim. Biophys. Acta, in press (1978)
175. Ikemoto, N.: J. Biol. Chem. *249*, 649–651 (1974)
176. Kalbitzer, H. R., Stehlik, D., Hasselbach, W.: Europ. J. Biochem. *82*, 245–255 (1978)
177. Froehlich, J. P., Taylor, E. W.: J. Biol. Chem. *250*, 2013–2021 (1975)
178. Kanazawa, T., Yamada, S., Yamamoto, T., Tonomura, T.: J. Biochem. *70*, 95–123 (1971)
179. Kanazawa, T.: J. Biol. Chem. *250*, 113–119 (1975)
180. Masuda, H., De Meis, L.: Biochemistry *12*, 4581–4585 (1973)
181. The, R., Hasselbach, W.: Europ. J. Biochem. *28*, 357–363 (1972)
182. De Meis, L.: J. Biol. Chem. *251*, 2055–2062 (1976)
183. Post, R. L., in: Biochem. of Membrane Transport. Semenza, G., and Carafoli, E. (Eds.) Berlin, Heidelberg, New York: Springer 1977, pp. 352–362
184. Whittam, R., Chipperfield, A. R.: Biochim. Biophys. Acta *415* 149–171 (1975)
185. Skou, J. C.: Quart. Rev. Biophys. *7*, 401–434 (1975)
186. Hasselbach, W., Makinose, M.: Biochem. Biophys. Res. Comm. *7*, 132–136 (1962)
187. Makinose, M.: Biochem. Z. *345*, 80–86 (1966)
188. Kanazawa, T., Yamada, S., Tonomura, Y.: J. Biochem. *68*, 593–595 (1970)

W. Hasselbach

189. Makinose, M.: 2. Int. Congr. Biophys. Wien 1966, 276, Verlag Med. Acad. Vienna
190. Yamamoto, T., Tonomura, Y.: J. Biochem. (Tokyo) *62*, 558–575 (1967)
191. Makinose, M.: Europ. J. Biochem. *10*, 74–82 (1969)
192. Sumida, M., Kanazawa, T., Tonomura, Y.: J. Biochem. *79*, 259–264 (1976)
193. Hasselbach, W., Makinose, M., in: Role of membranes in secretory processes. Bolis, L. (ed. *et al.*). Amsterdam: North Holland Publishing Comp. 1972, pp. 158–169
194. Makinose, M.: FEBS Lett. *25*, 113–115 (1972)
195. Beil, F. U., von Chak, D., Hasselbach, W.: Europ. J. Biochem. *81*, 151–164 (1977)
196. Kanazawa, T., Boyer, P. D.: J. Biol. Chem. *248*, 3163–3172 (1973)
197. Rauch, B., v. Chak, D., Hasselbach, W.: Z. Naturforsch. *32c*, 828–834 (1977)
198. Boyer, P. D., De Meis, L., Carvalho, M. G. C., Hackney, P. D.: Biochemistry *16*, 136–139 (1977)
199. Knowles, A. F., Racker, E.: J. Biol. Chem. *250*, 1949–1951 (1975)
200. De Meis, L.: personal communication (1977)
201. Rauch, B., v. Chak, D., Hasselbach, W.: FEBS Lett. *93*, 65–68 (1978)

Received April 18, 1978

Biochemical Aspects of Biomineralization

Gottfried Krampitz and Wiltrud Witt

Abteilung für Biochemie, Institut für Anatomie, Physiologie und Hygiene der Haustiere, Universität Bonn, 5300 Bonn 1, Germany

Table of Contents

1 Introduction

Biological calcification processes are widely distributed in nature. They can be found in microorganisms, in plants, in the animal kingdom and in humans. Under physiological conditions, the results of mineral deposition in biological systems can be represented by the formation of bones, teeth and shell material as well as coccoliths, corals, pearls etc. The variety of biomineralisates can best be expressed by the fact that approximately 128,000 species of molluscs[636] are known. The majority of them (Conchifera) form shells of different kinds of size and shape as well as of color.

There is, however, also a pathological aspect of biomineralization including the formation of body stones (kidney, gall-bladder), atherosclerosis, bone resorption and caries.

Biomineralization can be defined as a sequence of events whereby cells produce minerals which crystallize and add together according to a construction drawing. The results are so-called hard tissue or hard pieces. In spite of the variety of biological calcification products, all biomineralisates have one common feature. They are composed of an inorganic (mineralized) phase and an organic part which can be organized by proteins, phosphoproteins, glycoproteins, proteoglycans, polysaccharides, lipids etc. and frequently of mixtures thereof. Although at least twenty different skeletal minerals are reported to exist in living organisms[1, 2], only four are mainly found: aragonite, calcite, dahllite, hydroxyapatite, and opal[3]. The remaining minerals are either trace constituents or occur only in a few isolated species[3]. Crystals form by interaction of the organic phase with the inorganic substances. Our understanding of the molecular mechanisms of biological calcification processes are still superficial although many efforts from various sides have been made to shed more light on this unexplored territory of biochemistry. Most research so far in this field can be divided into three categories:

(a) morphological studies of the intricate relationships between mineral phase and organic macromolecule (matrix),

(b) investigations of the chemical composition of the mineralized tissue and the minerals, and

(c) exploration of ion transport mechanisms in cellular systems and the solid state principle involved in mineral deposition on organic substrates[3].

Biomineralization in its variety therefore is based on many classical disciplines of science, e. g. mineralogy, crystallography, palaeontology, geology, marine science, orthopedics, dentistry, cardiovascular science, urology, veterinary medicine, biology, agricultural sciences, nutrition, chemistry, biochemistry, physics, biophysics, physiology, and many more.

2 Components of Biomineralisates

2.1 Inorganic Components of Biomineralisates

Mineralogy, crystallography and chemistry of mineral deposits in living systems have frequently been reviewed in the past[4-44]. Therefore this will briefly be summarized here.

2.1.1 Calcium Carbonates

The following calcium carbonate deposits have been reported to occur in biological material:
calcite
aragonite
vaterite
$CaCO_3$ monohydrate "amorphous".
The two polymorphs of $CaCO_3$, e. g. the rhombohedral calcite (space group: R 3c) and the orthorhombic aragonite (space group: Pmen) are most common in organisms.

2.1.2 Calcium Phosphates

The biologically most relevant calcium phosphates are:
dahllite [carbonate hydroxyapatite, $(Na,Ca)_{10}(PO_4,CO_3)_6(OH)_2$]
francolite [carbonate fluoroapatite]
hydroxyapatite [$Ca_{10}(PO_4)_6(OH)_2$]
brushite [calcium monohydrogen phosphate dihydrate, $(CaHPO_4 \cdot 2H_2O)$]
octacalcium phosphate [$Ca_8H_2(PO_4)_6 \cdot 5H_2O$]
amorphous calcium phosphate [$Ca_3(PO_4)_2$?]
    While $CaCO_3$ crystals (calcite and aragonite) predominantly appear in egg shells and in biomineralisates from invertebrates, calcium phosphates are predominantly involved in processes which play an inportant role in medicine. They will be described here in detail; the knowledge of their structures are most relevant for the understanding of the cellular and molecular processes in bones and teeth.

2.1.2.1 Hydroxyapatite

Since no common agreement on the nature and mode of formation of calcium phosphates has been reached, an attempt will be made to coordinate the controversial observations[3].
    The inorganic phase of bones or teeth is mainly hydroxyapatite (HA), and deviation in Ca/P ratio from common HA (Ca/P = 1.667) is explained by the presence of amorphous phosphates[3]. The biogenetic HA resembles in size crystals of HA prepared by precipitation from aqueous solutions. The chemical composition of biominerals is similar to HA. However, crystals in bone, dentine and enamel can vary

greatly in size and shape[45−47]. An understanding of the structure of HA can be derived from the knowledge of the spatial organization of a small number of constituent ions. Possibly the whole crystal, regardless of size, can be generated (at least conceptionally) by translationally periodic repetitions of a basic structural pattern of the constituent ions known as the unit cell. The unit cell for HA is a right rhombic prism which, when stuck in the manner just described, forms a simple hexagonic lattice. The length along an edge of the basal plane of the cell is a = 9.432 Å and the height of the cell is c = 6.881 Å.

The spatial symmetry symbolized as $P\,6_3/m$, cannot be completely specified with less than this number of atoms. The arrangement of these constituent atoms are projected along the c-axis onto the basal plane. The hydroxyl ions lie, in projection, at the corners of the rhombic base of the unit cell. The hydroxyls occur at equidistant intervals one half of the height of the cell (3.44 Å) along columns perpendicular to the basal plane and parallel to the c-axis. Six of the ten calcium ions in the unit cell are associated with the hydroxyls. These calcium atoms form equilateral triangles centered on and perpendicular to the axis of the hydroxyl ions. Successive calcium triangles rotate 60° about this axis in a manner spaced one half a unit cell distance apart. The other four claciums of the unit cell lie along two separate columns parallel to the c-axis, heights halfway between the hydroxyl-associated calcium triangles. These "columnar" calcium ions are coordinated entirely by oxygens from the orthophosphate tetrahedra, which occupy the bulk of the space between the calcium ions in the structure[48, 49]. The hydroxyl ions do not lie in the center of the triangular planes defined by calcium ions[50]. The oxygen center of each hydroxyl ion is displaced by about 0.3 Å from the center of the nearest triangle of calciums. The hydroxyl ion is always perpendicular to the nearest plane of calciums with the hydrogen ion facing away from this plane in such a way that the O−H band never straddles the plane. In contrast, the fluoride ions in fluoroapatite (where chemically the fluoride replaces hydroxyl ions in apatite formula) lie in the center of the triangular calcium planes[51]. The closer coordination of fluoride as compared to hydroxyls, by the nearest calciums, may account in part for the greater chemical stability of fluoroapatite as compared to HA[51]. The view has been expressed that the axis through the triangular calciums is not an essential part of the main structure. Ions removed from these channels will not destroy the integrity of the structure if local charge balance can be maintained. In fact, in the apatite-like compounds $Pb_5(PO_4)_2\,SiO_4$[52] and $Ba_{10}(PO_4)_5PO_4$[53], these channels are indeed empty. Since the channels are large and are supported by the rest of structure, they provide easy diffusion paths by means of which fluoride ions may replace hydroxyl ions in fossil bone apatite[54]. In cation-deficient lead apatite, the empty cation sites were statistically distributed only along the columnar structural positions, and not among the triangular cation sites[55].

The dimensions of bone crystallites can be as small as 96 ± 10 Å, with the order of magnitude even a factor of two or three smaller[56]. The mean size of the apatite crystals in dentine and cementum is of the same order as that found in bone. Enamel crystals, however, are at least an order of magnitude larger in all dimensions. There appears to be good agreement that the smallest dimension of the bone apatite crystal is about 50 Å. On the other hand, there is a discrepancy in the reported size of the maximum length. From X-ray diffraction studies[56], it was derived that the largest dimension is probably less than 100 Å. This is consistent with the view of a bone apatite particle as a mosaic of microcrystals rather than as a continuously uniform single crystal.

Biological factors are also important in establishing the size, shape and orientation of bone crystals[57]. X-ray diffraction studies on ossified bird tendon show that the direction of the c-axis of the apatite crystallites was parallel to the collagen fibres[57]. In a low-angle diffraction study on compact bone[58], the apatite c-axis lies parallel to the largest bone crystallite dimension. This dimension coincided with the collagen fibre axis[58-61]. The side-by-side pattern, across the width of the fibril which appeared during the initial crystallization of the matrix, was followed by an end-to-end aggregation along the length of the fibril to produce the polycrystalline chains[59]. Since the c-axis of bone crystals grows parallel to the collagen fibre axis, fluoride arrangements suggest that the length of a bone apatite crystal in the direction of the c-axis is in some way limited by the length of the fundamental period of collagen and by the number of foci for crystal nucleation along this length. In enamel, where the matrix is said to be keratin(?)[62], no such restrictions in fluoride-induced growth along the c-axis occurs[63]. In the initial stage of the *in vivo* calcification of the collagenous matrix, the earlier mineral deposits are extremely small[64]. It has been concluded that the initial penetration of the matrix with mineral is quite rapid and is followed by a slower laying down of the mineral. Probably the slower step is a combination of growth of the initial seeds concomitant with the formation and growth of additional apatite, especially in the inter-collagenous spaces. Furthermore, much of the elongation of crystals associated with the collagen in more heavily mineralized areas is probably not a reflection of true crystal growth but of a "fusion" of the initial smaller crystals through separate growth of the latter[64].

Changes in the size of bone crystals with age have been noted in X-ray diffraction studies[65]. Here the mean size was observed to increase with age up to maturity. At this point a sharp leveling off of growth rate occurred and the size became relatively constant and independent of age. This could be considered as additional evidence for matrix-restricted crystal growth. However, it could also mean that secondary growth due to dissolution and reprecipitation does not occur in bone crystals in fully mineralized tissue. Secondary growth or ripening is encountered in aqueous systems of sparingly soluble salts in which the initial precipitate is finally divided[65].

Most of the HA crystals in young bone are small with a mean crosssectional width of only about 90 Å. Occasionally, larger crystals in the 500 to 1000 Å range have a thin platelet-like form. Those two crystal populations are morphologically distinct from one another and must have originated under somewhat different conditions[3, 48].

The major inorganic ions in bone mineral are calcium, phosphate and carbonate, with lesser amounts of magnesium, sodium, potassum, chloride, and fluoride[66]. Traces of iron, copper, lead, manganese, tin, aluminum, strontium, and boron have also been detected[67].

Considerable attention has been given to the site of these ions in bone salts with particular emphasis on structural interpretations based on the apatitic lattice model. Several problems, however, are encountered in any attempt to understand the stoichiometric structural relationships existing in bone mineral using such a model. One difficulty arises from the variability in the chemical composition of bone mineral itself. This variability is most noticeable in the levels of the minor constituents, but even calcium and phosphate fluctuate to some degree[67]. Also the Ca/P molar ratio ranged from 1.37 to 1.71. This ratio is age dependent, with the lowest values obtained for bones from children and the elderly and the highest for young adults[69].

The most challenging problems in defining the underlying structural basis for the observed chemical stoichiometry of bone mineral involve the two major constituents, calcium and phosphate. The HA model would predict that ideal composition for bone apatite should be $Ca_5(PO_4)_3OH$, with the calcium and phosphate in the Ca/P molar ratio of 1.67. Averaging data on mammalian bones result, however, in a Ca/P molar ratio of 1.74[67]. An average value of only 1.57, with a maximum value of 1.71 has been described[68]. The new Ca/P molar values would in every case be less than 1.67 leading to the assumption that bone HA is calcium deficient. Most conclusions concerning the structural origin of this calcium deficiency in bone apatites are based on studies of synthetic analogues. These explanations can also account for the reported wide variations in the magnitude of this deficiency[68].

Explanations for the existence of non-stoichiometric apatites on the basis of deficiencies in calcium ions can generally be divided into three groups:
(1) the calcium ions are missing from the crystal surface with replacement by hydronium ions, H_3O^+, to preserve charge neutrality[69],
(2) there are random defects in the apatite lattice due at least in part to statistical absences of calcium ions from interior lattice sites[55, 70],
(3) the calcium-deficient apatites are, in reality, a continuous series of lamellar intergrowth between octacalcium phosphate ($Ca_8H_2(PO_4)_6 \cdot 5 H_2O$) and HA[71].

None of the proposed models adequately take into account the possibility of both variations in surface composition and internal lattice defects. Studies of the crystallography of bones from young and adult humans demonstrate that bone minerals are not mono-mineralic but represent a mixture of several structurally different crystalline phases[48]. The major proportion of mineral species represent apatite which is best defined as an isomorphous mixture consisting of two end-members: HA and carbonate apatite, e. g. $Ca_{10}(PO_4)_6$ $(OH)_2$ and $(Na,Ca)_{10}$ $(PO_4,CO_3)_6$ $(OH)_2$. Furthermore, young bones contain calcium monohydrogen phosphate dihydrate (brushite, $CaHPO_4 \cdot 2 H_2O$)[48]. At pH 6.3 and about 40 °C the formation of octacalcium phosphate has been observed. This mineral has been reported from dental calculus.

Moreover, octacalcium phosphate ($Ca_8H_2(PO_4)_6 \cdot 5 H_2O$) has been found in young bones and dentine[72]; its possible role as a precursor of the small HA crystallites in bone and teeth has been discussed[73-79]. Boiling water and fluorine containing solutions readily hydrolyze octacalcium phosphate and convert it to HA. The therapeutic significance of fluorine has been reported[80]. The fluoride increases the crystallinity of the formed HA as measured by the broadening effects on the X-ray diffraction peaks.

Among the apatite substituents, carbonate has been the subject of many studies. The apatite present in dental enamel is not a pure HA, but rather a corbonate apatite with a carbonate content of 2–3 %. There is still a controversy about the location of the carbonate in enamel, dentine and bone. While most recent studies agree that carbonate appears to be substituting within the lattice rather than existing in an amorphous phase, there is still some disagreement as to the actual position of the carbonate. The presence and location of the carbonate in dental enamel may relate directly to the risk of carious attack. Carbonate has been shown to be leaked preferentially from early carious lesions[81].

In renal stones, apatite ($Ca_5(PO_4)_3 OH, 1/2 CO_3$) with a hexagonal shape has been observed; however, this mineral is difficult to identify because of its cryptocrystalline appearance. Struvite ($MgNH_4PO_4 \cdot 6 H_2O$) orthorhombic, newberyte ($MgHPO_4 \cdot 3 H_2O$) orthorhombic, whit-ockite ($Ca_3(PO_4)_2$) hexagonal, and also brushite have been reported to occur in renal stones[82].

2.1.2.2 Amorphous Calcium Phosphate

Among the possible apatite precursors, octacalcium phosphate and amorphous calcium phosphate (ACP) have been discussed. An ACP phase exists in bones along with a crystalline apatite phase. The presence of this phase and a quantitative evaluation of its percentage of the total bone mineral content have been determined by X-ray diffraction, infrared spectroscopy, and electron spin resonance technique[81]. The subject of amorphous phases in bio-phosphates is even more controversial than the nature of the crystalline phases[3, 111]. The round, doughnut-shaped appearance of the amorphous particles contrasts markedly with the straight-edge, solid neddle-shaped crystals of apatite seen in the same electron micrographs[83].

Unlike crystalline subsystems, one cannot understand the structure of an amorphous particle from a knowledge of the spatial organization of a small group of its constituent atoms. The reason is that the atoms are not arranged in a regular, periodic array which would enable one to define the whole space occupied by the particle by simple translational repetitions of a basic structural motif of atoms. The spatial organization of the ions comprising the amorphous material in bone mineral is, at present, completely unknown.

The ACP, like the synthetic apatites, do not have a rigidly defined chemical composition. The molar Ca/PO_4 ratios were found to vary from 1.44 to 1.55 depending upon the conditions of preparation[84]. On the other hand, the composition is less variable than that found in apatites prepared under the same conditions[85]. Despite the lack of constancy in the composition of these two materials, it was found that the converted apatite invariably had a higher molar Ca/PO_4 ratio than its amorphous precursor. In addition, the molar Ca/PO_4 ratio of the amorphous material having the greatest solution stability was found to be 1.50. This last observation is a possible chemical justification for considering the amorphous salts as non-crystalline tricalcium phosphates, rather than submicrocrystalline Ca deficient HA. The degree of instability of these amorphous materials would be related to the degree to which their composition deviates from the ideal tricalcium phosphate stoichiometry, $Ca_3(PO_4)_2$[84]. As far as is known, synthetic ACP preparations are not stable in aqueous media. If these phosphates are kept in contact with their preparative solution, these materials will hydrolyze into crystalline apatite. The rate of conversion is controlled by the apatite product and not by the amorphous precursor. In synthetic systems, this conversion rate has proved to be strictly proportional to the number of apatite crystals already formed and not to the amount of amorphous material remaining. Such autocatalytic behaviour is an example of secondary heterogeneous nucleation. Secondary nucleation phenomena can only occur in solutions supersaturated with ions of the crystallizing phase, in the present case with calcium phosphate and hydroxyl ions. The fact that the secondary mechanism is the only rate-controlling step in the conversion suggests further that the amorphous phase acts essentially as a passive, but very labile reservoir of calcium and phosphate, ensuring a steady state condition of supersaturation throughout the hydrolysis. The isothermal metastability of ACP in solution implies that the formation of this phase is a kinetic rather than a thermodynamic phenomenon[86].

ACP reveals discoidal and spherical forms[87], but it is questionable whether these two forms actually exist in living bone. It has been assumed to be more likely that they are produced by drying or aging of the sample materials. Amorphous mineral material occurs in matrix vesicles of newly synthesizing bone. This less dense amorphous mineral material appears to occupy the position of the vesicle membrane and it may be the precursor of bone material[88].

2.1.2.2.1 Stabilization of Amorphous Calcium Phosphate. It has been assumed by analogy to synthetic systems that ACP is a precursor phase which transforms autocatalytically to bone apatite unless stabilized in some way[89]. In line with this is the observation that young bone is higher in amorphous than in crystalline content, while the reverse is true for mature bone[78]. It appears to be the dominant phase in very early *in vivo* mineralization, and apparently persists in mature bone[78-92]. Because of its inherent instability and its continued presence in fully mineralized tissues, several studies have dealt with the stabilization of the ACP and with factors which might delay or prevent its conversion to HA[79, 93-97].

It has been shown that a variety of substances, including magnesium, pyrophosphate and a specific protein-polysaccharide aggregate, can in appropriate concentrations, delay the conversion of ACP to HA. Certain acidic phospholipids (PL) were bound to mineral in the early stages of *in vivo* mineralization[98-101]. The predominant one, phosphatidyl serine, has been found to be the major anionic PL in matrix vesicles associated with the new mineral formation *in vivo*[102, 103]. This lipid and other acidic PL are known to bind preferentially to calcium[104-107] and to have a stabilizing effect on ACP[96] in the presence of bone calcium and inorganic phosphate in biphasic systems[96]. Neutral zwitterion lipids which have amphiphatic properties but do not bind calcium are not effective in stabilizing ACP.

With preformed ACP at ratios of only one PL molecule per 30–50 Ca atoms, phosphatidyl serine markedly delays HA crystal formation. When phosphatidyl serine is present during ACP precipitation, inhibition of conversion to HA is less pronounced, but crystal habit and aggregation are greatly altered resulting in stacks of thin, membrane-like sheets approximately 3–42 Å thick. Phosphatidyl serine appears to be the most effective PL in blocking ACP to HA conversion, when oriented primarily on the surface. In view of the presence of anionic lipids in matrix vesicles and their association with early mineral deposits, these findings emphasize that lipids may play a role in the control of normal mineralization *in vivo*[108]. A calcium-PL-phosphate complex with a constant 1:1 calcium to total phosphate molar ratio has been shown to exist in bone[109]. This complex from young bone has been reported to initiate HA formation from metastable calcium phosphate solution.

The action of the complex is compared to that of the acidic PL: phosphatidyl serine, phosphatidyl inositol and phosphatidic acid. These PL first remove calcium, and a small amount of phosphate from the metastable solution, forming a material similar to the complex isolated from bone, and then form HA. The rate of HA proliferation, once phosphatidyl serine and phosphatidyl inositol are converted to $Ca-PL-PO_4$ complexes, is the same as the rate observed for comparable weights of the complex isolated from bone[110].

When calcium phosphate is precipitated from aqueous solutions of high supersaturation and pH values above 7, the solid phase appearing initially is an ACP with the formula $Ca_9(PO_4)_6$[111, 112]. If this amorphous precipitate is allowed to remain in contact with the solution, it transforms to crystalline HA through a process of dissolution, nucleation and crystal growth[77], unless stabilized in some manner. Under certain conditions, the mitochondria of many types of cells from a variety of animal species can accumulate large deposits of stable ACP[113]. In a tissue which calcifies, these mitochondrial granules might be involved in some aspects of the mineralizing process[113].

The mitochondria from the hepatopancreas of the blue crab contains ACP deposits that have been shown by X-ray radial distribution function analysis to be similar in local atomic order to synthetic ACP[114, 115]. In fact, ATP and Mg are required for the formation of this mitochondrial calcium phosphate[116]. Using concentrations of ATP in the same range as found in crab mitochondrial granules[115], it has been shown that the transformation of synthetic ACP to HA was considerably delayed. A much larger concentration of ADP was required to give the same transformation delay, but the effect was otherwise similar[117]. The presence of Mg in solution also delays the conversion of ACP to HA[118].

A synergistic effect has been demonstrated when Mg and ATP are used together in solution to delay the conversion of a slurry of ACP to crystalline HA. Conversion is delayed in some instances more than ten times as long as with either ATP or Mg alone. Conversion does not begin until ATP in solution is decreased through hydrolysis to an undetectable level. The effect of Mg is to decrease substantially the rate at which ATP hydrolysis occurs.

Once conversion begins, it proceeds more slowly in the presence of both Mg and ATP than with Mg or ATP alone. ATP was also found to prevent the formation of HA from metastable solutions of calcium and phosphate which did not contain any solid phase. Hydrolysis of ATP does occur in the presence of ACP or HA, presumably by transphosphorylation on the surface of the solid calcium phosphate phase. It was concluded that ATP stabilizes ACP, not by affecting its dissolution, but either by poisoning heteronuclear growth sites or by poisoning the growth of embryonic HA nuclei (formed heterogeneously or homogeneously) before their critical size is reached, or by poisoning both. In the case of embryonic HA nuclei, the poisoned nuclei would go back into solution preventing HA crystal formation. In addition, it was found that neutral $Ca_9(PO_4)_6$ clusters, which are believed to be the basic structural unit of ACP, break down into individual Ca and PO_4 ions when ACP dissolves in aqueous medium[119].

2.2 Organic Components of Biomineralisates

2.2.1 Lipids

It has been suggested that the acidic PL, phosphatidyl serine, phosphatidyl inositol and phosphatidic acid are involved in the initiation of biological calcification[120-123, 466]. The theory that PL are essential for the mineralization of vertebrate hard tissue[124] was initially advanced by histochemical studies and later supported by biochemical studies on animal models[99]. Independent studies of the lipid composition of hard tissues indicated that a specific fraction of the acidic PL found in dentine, enamel, calcified cartilage and bone could not be extracted by lipid solvents prior to demineralization[99, 125-127]. Evidence that the same PL fractions were necessary for HA formation came from the observation that the presence of these lipids is essential for the recalcification of the matrices of marmorset femurs[128] and dental calculus[129]. Furthermore, the PL extracted from bones, hard and soft tissues, were shown to cause the *in vitro* precipitation of HA from metastable solutions[130]. Although it was apparent that the acidic PL must in the same way be involved in mineralization, the precise function of the PL in tissue calcification and the nature of the interaction between PL and mineral remained uncertain. It has been suggested that the lipids might serve to stabilize ACP inhibiting the conversion of this amorphous precursor to apatite[108]. Others have suggested that a complex might exist between the PL, calcium phosphate and certain macromolecules and that this complex plays a role in the onset of calcification[96, 98, 101].

Phosphatidyl choline binds calcium, magnesium and rare earth cations. This has been studied by proton magnetic resonance and infrared spectroscopy. The calcium-induced chemical shifts for the various protons of phosphatidyl choline were C_α choline $> C_\beta$ choline $> N(CH_3)_3 > C_3$ glycerol. No significant chemical shifts were observed for the C_1 and C_2 glycerol protons. None of the acyl chain protons were affected by the presence of calcium. Analyses of the salt-induced shifts yielded binding curves which fitted excellently, with the theoretical data. The vicinal coupling constants for the various protons of phosphatidyl choline did not appear to change in the presence of calcium. Examination of the P=O stretching band $(1150-1300 \text{ cm}^{-1})$ of phosphatidyl cholines by differential IR spectroscopy showed that this band shifted to shorter wavelengths in the presence of calcium. The site of calcium binding to phosphatidyl cholines is deduced from proton magnetic resonance and IR data. It appears that both the structure and reactivity of the cation-lecithin complex are dependent on the nature of the cation[138].

The consistent finding of sizeable amounts of partially degraded PL in the matrix vesicle fraction[131-133], plus the demonstrated barrier to ^{45}Ca exchange imposed by lipid membranes, suggests that breakdown of the vesicle membrane may be an essential step to permit the enclosed ACP to convert to HA. Recent findings[134] have shown that the majority of the phosphatidyl serine and lysophosphatidyl serine present in the matrix vesicles is located on the inner side of the vesicle membrane. Thus, it is almost certain that the acidic $PL-Ca-PO_4$ complexes present are located on the inner surface of the vesicles. This would explain their inability to nucleate HA until the vesicle membrane was disrupted. With membrane rupture, two processes could occur:

(1) the complexes could be exposed to the extracellular fluid where it would be free to nucleate HA, and

(2) the ACP present in the vesicles could also convert to HA once the excess Mg present in the vesicle fluid was allowed to diffuse away. However, the mechanisms by which Ca and inorganic P enter the vesicle, and numerous other details, still remain to be established[135].

Lipids may also be involved in the calcification of the dentine matrix, because sudanophilic materials have been detected in ontoblasts, at the dentine-predentine junction and in the peritubular dentine matrix[311]. A series of nutritional studies have shown that dentine lipids are in a dynamic state[312−315]. A diet deficient in essential fatty acids produced degenerative effects on dentinogenesis and resulted in the production of teeth more susceptible to dental caries[315].

The major PL in predentine were phosphatidyl choline (52−56 %) of the total PL and ethanolamine phosphoglycerides (22−23 %). About 10 % of the ethanolamine phosphoglycerides were plasmalogens. The major fatty acids observed were palmitic, stearic, oleic, linoleic, and arachidonic acids. Predentine sphingomyelin contained only about 6 % of 24-carbon fatty acids, which is a relatively low amount compared to sphingomyelin in most other tissues[311]. No study of lipids in dental pulp, predentine or dentine has yet detected any change in the lipid composition during tooth development, even though lipids have been implicated in calcification[311]. The extractability of phosphatidyl serine and phosphatidyl inositol in dentine is markedly different from that of its precursor tissue predentine. Apparently all of the phosphatidyl serine and phosphatidyl inositol in dentine is strongly associated with calcium ions and cannot be extracted until the dentine is demineralized[311]. The fatty acid compositions of predentine and dentine are markedly different. Dentine and enamel of rats and dentine of rat and human teeth[316, 317] contain only a few percent of arachidonic acid and decosapolyenoic acids. Thus, the predentine with its complement of polyunsaturated fatty acids is apparently converted into a tissue dentine which contains essentially no arachidonic acid or decosapolyenoic acids[311].

In egg shells the presence of lipids has also been established and these have been characterized[136, 137].

2.2.2 Cyclic Nucleotides

It has been suggested[139] that adenosine 3',5'-monophosphate (cyclic AMP) functions as an intracellular "second messenger" which mediates the effects of specific hormones on their target cells. According to this concept, a hormone such as parathyroid hormone or calcitonin binds to a specific receptor site in the cell membrane where it activates adenyl cyclase. This enzyme accelerates the production of cyclic AMP (cAMP) from ATP in the cytosol. cAMP then activates enzymes which ultimately lead to cellular activities that are typical for the hormonal effect, such as bone resorption or apposition. Information regarding bone cell activation mechanisms is mainly derived from studies of hormonal effects on bone *in vitro*[140-146]. Since it is now well established that the basic mechanism of bone cell activation involves cAMP, it appears likely that this cyclic nucleotide will also have a determinant role in bone cell activation in response to mechanical stress.

Supportive evidence for this hypothesis was presented recently[144] in a study in which cAMP concentrations in alveolar bone extracts from orthodontically treated cats were found to be significantly higher than in non-stressed alveolar bone samples. A relationship has been suggested to exist between the cAMP-containing cells and the process of bone remodeling[147]. It has been shown that increased cAMP levels in cells are responsible for the activation of protein kinases[148] which catalyze the phosphorylation of other proteins and also stimulate calcium efflux from mitochondria. Both these effects of cAMP are considered to be important steps in the response of cells to external stimuli[149]. The observations that cAMP-containing cells are seen in areas in which cellular products such as osteoid are found relatively late, lead to the assumption that the presence of cAMP in cells means intracellular accumulation of substances such as lysosomal enzymes and other products which are to be released to the extracellular space at a later stage[150-152]. In view of these concepts which relate the presence of cAMP in cells to their response to external stimuli, it is conceivable that different degrees of cellular activity indicate different concentrations of cellular cAMP. Mechanical forces are capable of activating cells involved in bone remodeling, but this activation occurs in limited areas while most of the neighbouring osteocytes remain unaffected[147].

A considerable volume of experimental evidence supports the concept that cAMP serves as a second messenger in the action of parathyroid hormone (PTH) on bone cells[143-146, 153-156]. Recent reports indicate that cyclic guanosine 3', 5'-monophosphate (cGMP) may also serve as an intracellular second messenger in a number of tissues[157, 158]. However, little is known regarding this role of cGMP in bone. Recently it has been reported that bone samples taken from animals (female cats) 20 and 60 minutes after administration of parathyroid extract contained twice the amount of cAMP, and almost three times the amount of cGMP observed in the controls. These results indicate that the cellular source of bone cyclic nucleotides in parathyroid-extract-treated animals varies as to cell type, and therefore in bone the functions mediated by cAMP are not necessarily antagonistic to those mediated by cGMP[159].

2.2.3 Amino Acids and Peptides

Only very little information exists on the occurrence of free amino acids and peptides in biomineralisates. In extracts of dentine, four peptides have been identified (A, B, C, D)[160]. Peptides A and B contain those amino acids typical for collagen, peptide C lacks hydroxyproline, proline and phenyalanine, while peptide D is made up from lysine, hydroxylysine, glycine, alanine, arginine, glutamic acid and aspartic acid. Also a decapeptide has been isolated from dentine. Because of the content of three non-identified sugars, the latter peptide has been regarded as a glycopeptide[161-163]. Furthermore, a peptide has been found in dentine containing glycine, alanine, serine, glutamic acid, arginine and two unidentified components; the latter is assumed to be serine phosphate and γ-amino butyric acid.

Also free amino acids and peptides have been extracted from enamel. The material is very similar to that of dentine. Peptides from bone and teeth extract have an apparent molecular size of 750 to 5000 daltons[163]. Also peptides have been found in aqueous extracts of egg shells from birds and reptiles[164]. It has been assumed that free peptides and glycopeptides may have effects on the nucleation of the mineral phase[185]. On the other hand, those products could be split off from matrix proteins during crystallization processes. Perhaps there may be an association between free peptides and mucopolysaccharides or glycopeptides. It cannot be excluded that they are artefacts due to isolation procedures[164].

2.2.4 Citric Acid

Although about 80–90 percent of the total citric acid in humans are localized in hard tissues as enamel, dentine, cementum and bones, very little is known on the biological function of citric acid in biocalcification. HA crystals are reported to be dissolved by the action of citric acid. The acid dissolves the crystals in such a way that the destruction is a preferential attack along the c-axis. It is highly probable that the HA crystallites present in mineralized tissues also do have a dislocation in the centre of the material[165]. Another assumption describes that citric acid is a constituent of the aqueous phase of enamel or that citrate is bound to the surface of apatite by adsorption[166].

2.2.5 Organic Matrix Components

Biochemical and morphological studies have shown that at least matrix components and matrix vesicles are closely involved in biomineralization. The concept of an organic matrix in any hard tissue depends on the hypothesis, that macromolecular organic constituents largely control the deposition of solid inorganic matter. The term "matrix" is widely employed in a histological context but its popularity has resulted in frequent laxity of usage through reference to any continuum surrounding discrete elements as a matrix. For example, the "extracellular matrix of cartilage" is no matrix for the cells because it is produced by them, not vice versa. More precise application of the terms is easier in relation to inorganic crystallites in mineralized tissues. Its derivation from *mater*, through the Latin *matrix* (womb or uterus) emphasizes the main features of a general definition in current English as "the place or medium in which something is bred or developed". This provides the criteria applying to a true histological matrix, which should:
(1) be in existence before whatever is bred
(2) somehow participate in its development
(3) subsequently enclose it spatially.

Matrices of mineralized tissues may be subdivided[167] into (a) permanent ones which (4a) remain within the tissue, continuing to enclose the inorganic phase (e. g. mineralized keratins, bone, cementum, shells and dentine) and (b) temporary matrices which (4b) eventually leave the inorganic crystallites in a comparatively unprotected state, exposed to the influence of the external environment (e. g. dental enamel).

Concerning the chemical nature of the matrices of mineralized tissues, macro-molecular polyions invariably seem to be present. These are usually predominantly proteins. Occasionally the matrix is rich in highly basic polyanions and polyuronic acids. The chemistry of organic matrices from invertebrate animals has received little attention until recently when a brachiopod exoskeleton which contains apatite has been investigated[168]. The matrices of 8 species from 4 phyla have been studied in which either silica or calcium carbonate forms the main inorganic phase[169]. In two species of echinoderms the main matrix constituent was a collagen of characteristic composition and striated fibrillar structure which, however, gave rise to calcite crystals in contrast to vertebrate collagens which form apatite. Collagen was a minor constituent of the matrices of mineralized tissues from representatives of the phyla Porifera, Mollusca and Arthropoda. Other matrices were formed by conchiolin (molluscs) and by proteins and protein-complexes (egg-shells). Extracellular matrix, or briefly matrix, is defined as anorganized solid or semisolid extracellular material on the surface of individual cells and on the surface or in between cells of multicellular organisms. The assembly, organization, and specialization of cells in multicellular systems, however, occur only when the intercellular matrix develops, protecting, connecting, binding, and organizing the cells into one functional system of tissue, organ or organism. This environment may play a primary role in the control of cell processes. Thus, the cellular function of the multicellular organism cannot be fully understood without complete knowledge of the structure and function of the matrix.

A variety of organic matrices are assumed to be able to initiate or accelerate the precipitation of calcium and phosphate ions from solutions[171–177], by a process called heterogeneous nucleation[178–180]. Since mineral is normally deposited extracellularly, these observations may help to explain the manner in which the mineral phase of bone is formed and localized. Many of these studies were carried out with collagenous matrices in which the collagen fibres are highly ordered[176, 181, 182]. Therefore, it has been proposed in analogy with nucleation phenomena in inorganic chemistry that such organization is a necessary characteristic of a template which can influence the rate of crystal formation[177–179, 183].
Interest in the role of the matrix in mineralization was reawakened by the demonstration, that the crystals in bone were highly oriented with respect to the collagen fibres. This in turn has led to studies on the chemical nature of the matrix related to its ability to calcify[184].

By contrast and despite many studies, the role of the components of the calcifying matrices (collagen, fibrils, proteoglycans, lipids, phosphoproteins) is still uncertain. This material is thought to play an essential role in inducing crystal formation[77].

2.2.5.1 Collagen

Collagen is the major protein of the extracellular connective tissues and functions as a structural protein for many biomineralization processes, e. g. in bone and teeth. The amount of collagen varies from one species to another and from tissue to tissue within the same species. Collagen comprises as much as one third of the total protein in the vertebrate body and in certain invertebrates, such as sponges and echinoderms as well as corals[186], collagen may account for an even larger proportion of the total protein of the organism.

Collagen has long rod-like molecules consisting of three polypeptide chains coiled left-hand about a common axis[187, 188]. The dimensions of about 3000 x 15 Å are well established[189]. Chemical evidence that the major vertebrate collagens contain three similar (but not necessarily identical) chains extending in parallel the full length of the molecule has been summarized[190]. These are designated α-chains. Such chains predominantly have a molecular weight of approximately 95.000 daltons[193–195]. In general the amino acid composition of one of the three chains (α_2) differs significantly from that of the other two (α_1)[195–198]. Although some variations have been noted in collagen composition, about one third of the total amino acid content is glycine and about 20 % are amino acids. Most vertebrate collagens contain three uncommon amino acids: 4-hydroxyproline, 3-hydroxyproline and 5-hydroxylysine. These proteins lack cysteine and are low in aromatic amino acids[184].

The introduction of covalent cross-links produces double-chain β-components, triple-chain γ-components and higher molecular weight products which can be identified in solutions of denatured collagen[190]. Denaturation of collagen results in the appearance of three molecular components, termed α-, β- and γ-components. The molecular weight of the α-component is considered to represent any single chain of which there are three in the collagen molecule. By chromatographic techniques the components from the major portion of vertebrate collagen, including those from dentine and cementum, are separable into two distinct units which have approximately the same molecular weight (100,000). They are present in a ratio of 2:1. The chain composition is $[\alpha_1]_2\alpha_2$. Recently collagens with α-chains with amino acid compositions different from α_1 and α_2 have been described. Arbitrarily, these α-chains have been designated α_1 in conjunction with a Roman numeral designating their sequence of discovery. Previously described α-chains with amino acid compositions different from α_1 have retained their original nomenclature, e. g. α_2 and α_3. Thus, the first described and predominant collagen type contains α_1 (I) and α_2 type chains and has a chain composition of $[\alpha_1(I)]_2\alpha_2$. A minor but significant portion of vertebrate collagens have chain compsitions different from $[\alpha_1(I)]_2\alpha_2$. For example, cartilage collagen is composed almost entireyl of molecules containing three identical chains with respect to amino acid composition and sequence. These chains are designated α_1 (II) and the molecular chain composition is designated $[\alpha_1(II)]_3$. Other vertebrate tissues containing genetically distinct α-chains and collagen molecules with three identical chains are skin and basement membrane. The β-components have a molecular weight of 200,000 daltons and an amino acid composition indicating they are dimers consisting of two α-chains. These are formed by covalent cross-links between α-chains. γ-components have the same amino acid composition as that of unfractionated tropocollagen and probably represent many possible combinations of trimers of covalently linked α-chains[191].

Significant advances have been made in elucidating the primary structure of α-chains. The amino acid sequence of the α_1- and α_2-chains of the helical region of collagen, residues 1–393 have been already reported[192].

Almost the complete amino acid sequence of an α_1(I)-chain (more than 1000 amino acid residues) has been determined[191]. From this work, this collagen α-chain may be defined as a linear sequence of amino acids, linked only by α-amino, α-carboxyl peptide bonds. The α-chains are characterized by glycine at every third position for more than 95 % of their length. Amino- and carboxy-terminal ends of collagen α-chains do not contain the repetitive glycine structure. Therefore, short sequences (16 or 17 residues) called telopeptides, which differ in structure from the rest of the collagen molecule in that they are probably devoid of the polyproline kind of the helical structure, are present at the terminal ends of the α-chains. The N-terminal telopeptide region of the α-chains contain the specific lysine and hydroxylysine residues that eventually take part in intramolecular cross-link formation. Also the N-terminal region contains part of the immunogenic determinants of the collagen molecule.

It has been known for a long time that collagen is a glycoprotein. The carbohydrate is glycosidically linked to hydroxylysine as 2-O-α-D-glycosyl-O-β-D-galactosylhydroxylysine and O-β-D-galactosylhydroxylysine[199−201]. The role of the carbohydrate in collagen has been much debated. The presence of large amounts of carbohydrate in collagen from tissues such as cartilage suggests that the carbohydrate may be important in interactions between mucopolysaccharide and collagen. Just how this may occur, however, is not at all clear[187].

The insolubility of mature normal collagens has been ascribed to the presence of a system of covalent intermolecular cross-linkages that, in effect, convert a collagen fibre into an infinite cross-linked network of its monomeric elements. It has been concluded that at least two sets of cross-linkages of different chemical character exist, one set within structures which could best be represented as long filaments while the other set are bands transverse to the filaments and joining them together[202]. The formation of stable intermolecular cross-links in biological macromolecules correlates intimately with growth and development. Many of these processes involve formation, maturation, and aging of the fibrous structural proteins[203]. The cross-links are derived from protein-bound lysine and hydroxylysine and their aldehyde derivatives allysine and hydroxyallysine. Cross-linking may be conceived as forming between chains through an aldol condensation reaction between allysyl or hydroxyallysyl residues or as Schiff bases arising from lysine or hydroxylysine and their aldehyde derivatives.

Several types of cross-links have been isolated from collagens derived from various tissues. Depending on the participating residues, the initial condensation product and the subsequent chemical modifications, collagens may contain any of a variety of cross-links. A major type of cross-link in highly insoluble collagen of calcified structures, such as bone, dentine and cartilage, has been identified as dehydrodihydroxylysinonorleucine and its derivatives. It is highly probable that dehydrodihydroxylysinonorleucine is present in tissues in the reduced form (dihydroxylysinonorleucine) or as the keto-amine derivative. Such stabilized forms of dehydro-dihydroxylysinonorleucine have been assumed to be the residues responsible for the insolubility of collagen from mineralized tissues.

The absence of half-cystine residues in collagens with chain compositions $[\alpha_1(I)]_2\alpha_2$ and $[\alpha_2(II)]_3$ exclude cystine disulfide bridges from participation in cross-linking in these collagens. However, half-cystine residues have been identified in $[\alpha_1(III)]_3$ collagen and disulfide bridges may serve as cross-links in this type of collagen[191].

Considering minor departures from normal in the composition of collagen in the pathological bone, proline and hydroxyproline and the amino acid with small side chains, glycine and alanine, are all reduced by 7—9 % in osteogenesis imperfecta. The overall deficiency in osteogenesis imperfecta of these amino acids present in large amounts of collagen, together with an excess of amino acids with lipophilic side chains, indicates the possible presence of non-collagenous protein (possibly glyco-protein) which is more difficult to remove than from normal bone collagen. A possible alternative explanation is that, in osteogenesis imperfecta, the extra-cellular organization of some collagen may be arrested at an early stage[204].

The number of lysyl residues of bone collagen obtained from rats with experimental osteoporosis are significantly larger than those of the control group. However, the hydroxylysyl residues showed no difference between them. This suggests the decrease of cross-linking of collagen with osteoporotic bony change[205]. The amino acid composition and cross-linkage pattern of collagen extracted from osteo-arthritic bones was similar to that of young tissue[206].

Fresh bone contains cellular proteins and a certain amount of glycoprotein which contributes glucosamine to an analysis, but most protein in even a fresh bone or tooth is collagen.

Collagen of fossil bones has much the same composition no matter what the source[214] is. Though the amino acid composition is always that of unaltered collagen, the quantity of protein preserved in these fossils varies greatly from specimen to specimen. Minor differences do exist in the amino acid composition of fossil collagenous residues of the tables but it is hard to know how much significance should be attached to them. In spite of the close similarity between the collagens from all living animals, there are real but small differences and it will be important to ascertain if fossil collagen also show them. Dinosaur collagens have contained practically all the amino acids present in younger fossils except for reduced amounts of the relatively less stable hydroxyl- and sulfur-containing acids[214].

The decalcified matrix of human and bovine *dentine* is composed essentially of collagen. Dental protein possesses the distinctive wide-angle X-ray diffraction pattern and amino acid composition of collagen. No significant difference in amino acid composition of dentine is observed as the tooth matures[191]. Dentine collagen has a high degree of insolubility[299—301]. In addition, the dentine matrix collagens show little detectable swelling in acid, and do not appear to either swell or became denatured in strong hydrogen-bond breaking agents[299]. These properties can be attributed to a combination of two factors: the mechanical, strengthening of the tissue as a result of the very tight fibre weave and a high degree of intermolecular cross-linking[302]. In dentine collagen, the major intermolecular cross-link is derived from the condensation of hydrolysine-aldehyde and the ϵ-amino group of hydroxy-lysine from an adjacent molecule. The resultant aldimine bond then spontaneously untergoes an Amadori rearrangement to form stable cross-links (hydroxy-lysino-5-keto-norleucine). With increasing age, most collagenous tissues, e. g. skin, tendon, and bone, exhibit a decrease in the ratio of dihydroxy to monohydroxy cross-link from fetal to adult tissue, followed by gradual reduction in the total content of reducible cross-links during maturation. These changes however, do not occur in dentine collagen, leading to the suggestion that the maturation process of this collagen is different[303]. β-aminopropionitrile had little or no effect on the solubility of dentine collagen[304].

Dental *cementum* is a tissue of highly specialized functions, serving as anchorage for the collagenous fibre bundles of the periodontal ligament. Despite the importance of cementum for the dental attachment apparatus, only limited knowledge exists of its composition and metabolism[318]. Previous investigations have shown that collagen is the main component of organic matrix of cementum[319, 320]. This collagen has further been characterized as being mainly type I with the chain composition, $[\alpha_1(I)]_2\alpha_2$[321] and thus similar to the predominant type of collagen found in bone and skin[322, 323], type I collagen $[\alpha_1(I)]_2\alpha_2$ accounts for more than 90% of the organic matrix while type III collagen $[\alpha_1(III)]_3$[324] is present at a level of approximately 5%. Amino acid analyses revealed that the CNBr peptides from $[\alpha_1(I)]$ chains of cementum closely resemble the corresponding peptides from calf skin. The only systematic difference is a higher level of hydroxylation of prolyl- and lysyl-residues of the cementum peptides[318]. Human cementum consists of type I collagen only as identified by amino acid and hexose analyses[325].

2.2.5.2 Non-Collagenous Polymers in Bone

The non-collagenous proteins of bone comprise approximately 10% by weight of the organic matrix[210]. However, only 40–65% of the non-collagenous fraction is readily extractable and recoveries are variable[215], although improvements of the extraction techniques have been developed very recently[216]. In bone, various proteins other than collagen have been identified. Among them are those of protein synthesis[207], enzymes of protein and phosphate catabolism[208], alkaline and acid phosphatases[209]. In comparison with cartilagenous tissues, there is a paucity of protein-polysaccharides but relatively large proportions of peptides[211].

Preliminary experiments indicate that several non-collagenous proteins, even with Ca-binding properties, exist in bovine bone[217].

2.2.5.2.1 Calcium-Binding Proteins. In the non-collagenous protein fraction extractable from bone, a calcium-binding protein has been isolated that contains three γ-carboxy glutamic acid (Gla) residues with a molecular weight of 5700 daltons. This protein (Gla-protein) from calf cortical bone accounts for 1–2% of the total protein in calf bone. A similar Gla-protein has been found in swordfish vertebrae and human tibia as well as in chicken tibia[638]. The abundance of this protein (osteocalcine[637]) and its presence in all vertebrate bones which have been examined indicate that it has an important function in calcified tissues. Also sites of pathological ectopic calcification, e. g. renal stones, atherosclerotic plaques, contain Gla-protein[638]. This protein strongly binds to HA crystals but not to ACP, and inhibits HA crystallization from supersaturated solution. These properties of the Gla-protein may be conferred upon it by Gla, in analogy with the function of this vitamin K-dependent amino acid in $Ca^{2\pm}$ and $BaSO_4$-binding properties of the blood coagulation factors.

The partial amino acid sequence of Gla-containing proteins from calf bone and swordfish bone shows that 21 of the 45 residues common to each are identical. The three Gla-residues and the single disulfide bond are in identical relative positions in the two proteins. An important difference between both proteins is the absence of 4-hydroxyproline in the swordfish protein[638]. The kidney has been localized as the site of vitamin K-dependent Gla-protein biosynthesis[637]. Gla has been reported to be a constituent of fossil bones[639].

2.2.5.2.2 Phosphatases. Of particular importance is the observation that enzymes obviously involved in biomineralization processes occur in bone.

Evidence for the existence of two or more acidic phosphatases in bone was obtained with electrophoresis and assay with α-naphthyl phosphate[218]. However, bone alkaline and acid phosphatases have partially been seperated on Sephadex G-200[219], but showed only one acid phosphatase peak. At least two different enzymes[220, 221] in bone have been found to hydrolyze the substrates β-glycerophosphate and inorganic pyrophosphate: (1) a tartrate-sensitive enzyme with high specifity for β-glycerophosphate and (2) a tartrate-resistant enzyme with high specifity for pyrophosphate and p-nitrophenyl phosphate. Bone extracts have been chromatographed on CM-cellulose[222]. Most of the acid phosphatase activity was bound to the cellulose and could be eluted as one distinct peak. The major peak was strongly bound to the cellulose and showed high activity with p-nitrophenyl phosphate and inorganic pyrophosphate, but only slight activity with β-glycerophosphate and was unaffected by tartrate. The minor peak was weakly bound to the adsorbent, showed equal activity with p-nitrophenyl phosphate and β-glycerophosphate, but negligible activity with pyrophosphate and was completely inhibited by tartrate. Bone contains at least two different acid phosphatases and the more abundant enzyme may function as a pyrophosphate. The physiological substrate for the latter enzyme may be pyrophosphate or another oligophosphate ester. The first enzyme may utilize only monophosphate esters.

Studies of acid and alkaline phosphatase activities in bone have been used previously to investigate the action of various substances such as parathormone[223, 224], calcitonin[225], diphosphonates[226] and vitamin A[227]. Calvarial acid and alkaline phosphatase activities are modified after in vivo injections of vitamin D_3 metabolites to rats fed a vitamin D-deficient, low-calcium (0.02 %)diet. Changes in these activities are observed after administration of small doses, i. e. 130 pmol of $25-(OH)D_3$, $24.25-(OH)_2D_3$ or $1.25-(OH)_2D_3$. Each of these active metabolites has a different effect on bone phosphatase activity:

(a) A single injection of $25-(OH)D_3$ promotes an increase both in acid and alkaline phosphatase contents of rat calvaria. After one week of daily injections these changes persist except for acid β-glycerophosphatase;

(b) The administration of one dose, as well as of seven daily doses of $24.25-(OH)_2D_3$, is followed by a significant decrease in all enzymatic activities;

(c) $1.25-(OH)_2D_3$ administration induces a decrease in acid and alkaline phosphatase activities 24 h after the first injection. After seven injections the only effect observed is a decrease in β-glycerophosphate activity.

The changes in calvarial phosphatase activities observed in animals treated with 25–(OH)D$_3$ are totally different from those obtained with either 1.25–(OH)$_2$D$_3$ or 24.25–(OH)$_2$D$_3$. This fact indicates that physiological doses of 25–(OH)D$_3$ may have an effect on cellular activity, independent of the conversion of this metabolite into these dihydroxyderivatives. The various effects of these vitamin D$_3$ metabolites cannot be correlated with changes in serum calcium and/or phosphate concentrations. Among those factors other than serum calcium and phosphate concentrations that may be involved in the mechanism of action of vitamin D$_3$ metabolites on bone phosphatase activities, the parathyroid hormone is of importance. This hormone is known to be a potent activator of bone phosphatases[223, 224, 228]. Parathormone increases the content of alkaline, neutral and acid phosphatases in mouse calvaria in vitro. Calcitonin does not prevent the increase of those enzymes while dichloromethylene diphosphonate causes a decrease in acid phosphatase and pyrophosphatase[226].

The non-specific alkaline phosphatases present in bone and calcifying cartilage have several properties in common. The ATPases concerned in the formation of different hard tissues seem to be isozymes. It could be shown that two enzymes capable of degrading ATP exist. One of them can be inhibited by levamisole and R 8231 and is probably identical with non-specific alkaline phosphates. The activity of the other enzyme, tentatively named "Ca-ATPase", is dependent on the presence of Ca^{2+} or Mg^{2+} and is activated by these ions. The "Ca-ATPase" is unaffected by ouabain and ruthenium red. It may be speculated that the "Ca-ATPase" is concerned with the transmembranous transport of Ca^{2+}-ions to the mineralization front[229].

2.2.5.2.3 Glycoproteins. Extraction of non-collagenous proteins from bovine bone yielded three fractions, including a sialoprotein[212, 213], which are distinctly different in their chemical composition and in their electrophoretic mobility on cellulose acetate. A fourth fraction, consisting of material insoluble in acetate buffer at pH 5.0, contained a high proportion of collagen and apparently proportions of the other fractions. Sheep cortical bone matrix contains several glycoproteins which can be fractionated[242]. An acidic glycoprotein from bone matrix may be bound to collagen[242, 243]. In sheep cortical bone matrix, a sialic acid-rich glycoprotein in association with chondroitin sulfate was observed[243] whose composition is very similar to that of the chondroitin sulfate-sialoprotein complexes isolated from bovine extracts[244]. Two sialoprotein fractions have been isolated from bovine cortical bone[245]. One is a homogeneous glycoprotein which has been purified and studied in some detail[246, 247]. The second is the above mentioned sialoprotein associated with chondroitin sulfate[244].

Extracts of sheep bone contain small amounts of serum albumin and immunoglobulin-G[242].

Plasma proteins have been shown to be incorporated into normal bone matrix[230, 231, 238]. One plasma glycoprotein, identified in the human as the α_2-HS-glycoprotein, is concentrated in bone and permanent dentine between 30–100 times in comparison with other plasma proteins. The α_2-HS-glycoprotein is also concentrated in comparison with serum albumin in the mineral phase following either the precipitation of apatite in serum samples or the addition of either apatite or ACP to serum. The α_2-HS-glycoprotein binds calcium (one binding site for calcium per molecule); however, this protein was unable to nucleate calcium phosphate deposits from associates with mineralized tissues and it is suggested that the enrichment of the α_2-HS-glycoprotein in bone is a result of its preferential uptake from the bone tissue fluid by the developing mineral[232].

A similar or possibly the same glycoprotein of α-electrophoretic mobility which can be isolated from bovine and rabbit bone matrix is present in the blood plasma[233, 234]. The α-glycoprotein is present in a relatively high concentration in plasma and the majority of the plasma proteins of quantitative importance are manufactured by the liver[235]. There is strong support that the α-glycoprotein is taken up by bone from the plasma during bone formation[234, 236].

The extractability of [131]J-labelled plasma albumin from bone pieces and from powdered bone has been compared after both in vivo and in vitro incorporation. Albumin is more readily extracted from bone pieces than from bone powder which implies that tissue disruption exposes additional protein adsorption sites. It has been suggested that incorporation of plasma albumin and other proteins into calcified matrix during bone formation occurs mainly as a result of its strong interaction with bone mineral[237, 239, 240].

2.2.5.2.4 Proteoglycans. Proteoglycans are directly implicated in the process of calcification. In cartilage, approaching the calcification front, the amount of proteoglycans increases in parallel with dramatic decrease of collagen. A glycoprotein specifically present in calcifying cartilage shows high affinity for calcium ($K_d = 10^{-7}$ M) and also displays alkaline phosphatase activity. ATP and pyrophosphate substrates, highly implicated in the process of calcification, are also hydrolyzed. Proteoglycans from nasal septum are more efficient than those from preosseous cartilage in preventing calcium phosphate precipitation "in vivo". The molecular organization of proteoglycans is different in a calcifying tissue as compared to that of a non-calcifying one. Proteoglycans extracted from resting cartilage as well as those from the non-calcifiable nasal septum show high degree of aggregation. Conversely, proteoglycans from ossifying cartilage are not aggregated. Hyaluronic acid, which promotes aggregation of proteoglycans extracted from the former tissues, does not react with proteoglyeans from the latter type of cartilage. Collagen favours, however, aggregation of these molecules whereas it has no effect on proteoglycans from nasal septum and a slight effect on those from resting cartilage. Early calcification would then occur in a microenvironment of disaggregated proteoglycans, where Ca^{2+} is bound to a glycoprotein able to split phosphate from pyrophosphate, a known inhibitor of calcification[241].

2.2.5.2.5 Glycosaminoglycans. Several reports[248−260] have pointed out that glycosaminoglycans which occur in tissues in the form of proteoglycans may play an important role in calcification processes. In endochondral ossification, the proteoglycans and their glycosaminoglycan components are considered to be of importance in the sequence of biochemical processes which characterize the transformation of epiphyseal cartilage into bone[259]. Epiphyseal cartilage contains large amounts of glycosaminoglycans while the adjacent newly-formed bone shows an abrupt reduction in glycosaminoglycan concentration[261]. Predominantly chondroitin-4-sulfate is present in normal epiphyseal cartilage and chondroitin-6-sulfate in the case of the rachitic cartilage[262]. Unextractable glycosaminoglycans in hypertrophic and calcified zones are mainly of lower molecular weight and/or charge density compared to the extractable pool. Hyaluronic acid was inextractable in resting columnar and hypertrophic zones with increasing concentrations towards the calcification front. In calcified zones a shift to mainly extractable hyaluronic acid occurs[259]. Intracellular acid glycosaminoglycans of isolated chondrocytes from metaphyseal zones of embryonic chick bones are of short chain length and low charge. This is not the situation in normal cartilage where sulfation appears to be completed intracellularly[259, 263]. The change in the population of intracellular glycosaminoglycans in isolated chondrocytes is the result of highly increased synthesis compensating the loss of the surrounding matrix components and partly the release of originally synthesized glycosaminoglycans into the medium[264].

Ethane-1-hydroxy-1,1-diphosphonate has a general inhibitory effect on the synthesis of various sulfated glycosaminoglycans, which caused no qualitative change in the glycosaminoglycan.

In congenital pseudarthrosis, chondroitin sulfate is the major component in the fibrous, the cartilagenous and the osseous area; a significant amount of dermatan sulfate occurs in the fibrous region. A small amount of hyaluronic acid has been found in both fibrous and osseous areas[257]. Iliac crest cartilage from patients with spondyloepiphyseal dysplasia congenita contains increased amounts of chondroitin-6-sulfate and keratin sulfate[265].

2.2.5.3 Non-Collagenous Polymers in Teeth

The organic matrix of developing immature dental *enamel* consists of a class of proteins and polypeptides whose composition[179, 266−271] and molecular structure[272−274] distinguish them from collagen, the major structure protein of bone and dentine. The immature enamel proteins exist in aqueous solution as a multi-component system of rapidly interacting protein aggregates[266, 268, 275, 276−279], with molecular weights as great as $2-3 \times 10^6$ daltons[280]. However, there is evidence for the conclusion that the organic matrix of immature bovine enamel consists primarily of a large number of relatively low molecular weight components (15,000 daltons) up to a stage of development where the matrix can still be dissected as a tissue fabric after decalcification and before a massive loss of proteins occurs. The immature enamel matrix represents a mixture of proteins, ranging from the most immature enamel matrix to an enamel matrix which has developed to the point where it has been impregnated with considerable amount of the inorganic mineral phase. The soluble proteins of this grossly immature enamel matrix therefore represent constituents which may already have undergone to varying degrees some of the changes occurring in development.

If the ameloblasts initially synthesize high molecular weight species as suggested[281−285], then those observations imply[280] that degradation of these components occurs in vivo and presumably gives rise to the many low molecular weight components. Moreover, the absence of significant amounts of high molecular weight species in the immature embryonic enamel at stages where the organic matrix still retains it fabric-like structure in EDTA, with its major protein constituents insoluble in EDTA, suggests that the degradation of high molecular weight components occurs very rapidly after their synthesis and precedes the massive loss of protein components which occurs during the final stages of calcification and maturation[286-289]. The most immature enamel contains about 15–20 % protein by weight, whereas the most mature enamel contains 0.1 % protein or less. The proteins in the organic matrix of immature enamel are relatively insoluble in nearly neutral solutions of 0.5 M EDTA and are characterized by their high concentrations of proline, glutamic acid, leucine, and histidine ("immature" protein components). Conversely, the proteins and peptides of fully mature enamel are soluble in 0.5 M EDTA and are characterized by their high concentrations of aspartic acid, glutamic acids, serine, and glycine ("mature" protein and peptide components). The very marked decrease in the protein content of maturing and mature enamel and the differences in amino acid composition of mature enamel compared with immature enamel, suggest that maturation of enamel is biochemically characterized by the selective loss of certain peptides and the retention of certain other peptides, the latter possibly due to their interaction with the mineral ion constituents[290].

The biochemical processes accompanying maturation appear to begin early in the development of the enamel at a time when the organic matrix and its protein constituents are immature. If consists of the selective loss and the selective retention of certain components from the enamel suggesting protein and/or peptide degradation or depolymerization.

There is no evidence that the overall compositional changes in the enamel at any stage of development is principally due to compositional changes in the individual protein and peptide component *per se.* It is proposed that if peptide bond hydrolysis occurs during enamel maturation, it is catalyzed in part by the enamel crystals[290].

The most obvious changes in protein content and amino acid composition occur in the vicinity of a white opaque band[291]. The existence of enamel matrix proteins in an incompletely mineralized form in the area generally described as amorphous cementum, is consistent with the findings of enamel protein on the root analogue surfaces of rabbit incisor teeth. Comparison of the electrophoretic profiles of labial and lingual protein extracts shows that there is no qualitative difference, indicating that these two sides of the tooth are composed of identical proteins at this stage of development. Three proteins identified as enamel proteins[213] stain much less heavily on lingual gels as compared with labial gels[292, 293]. Little is known about the chemical nature of nucleation centers in enamel[170], but serine phosphate in enamel proteins has been identified[295] and considered probably to be implicated in the mineralization process[170]. Phosphorylation of serine residues in these proteins can result from the action of protein phosphokinase from muscle[296].

Since almost all of the organic weight of immature[297] enamel consists of protein[266, 268-271, 274, 276, 294], the presence of minor organic substances, e. g. lipids and carbohydrates, may be neglected[298].

The total non-collagenous matrix of dentine has been separated into two major fractions, a less acidic fraction A and an anionic fraction B by ion exchange chromatography. The less acidic fraction has been further resolved by isotachophoresis; twelve fractions were obtained. Each contained approximately 20 percent of carbohydrate, mainly in the form of neutral hexoses, including fucose. The molecular weights were in the range of 12,000 to 15,000 daltons except for one which was 26,000 daltons. The anionic fraction has been separated into three anionic glyco-proteins. Two of these (mol. weight 22,000 daltons) contained 1.80 % phosphorous and 4.16 % sialic acid and small amounts of hexosamines. α_2-HS-glycoprotein was identified as a component of fraction A[305]. Phosphoproteins from dentine of various species have been studied[306–308]. Very recently two phosphoproteins with molecular weights of 71,000 and 65,000 daltons, respectively, have been isola-ted[640]. Those phosphoproteins differ in content of apolar amino acids, although both contain >70 % of seryl (or phosphoseryl) and aspartyl residues. The name "phosphophoryns" has been proposed to describe these dental proteins. Two col-lagen-phosphoprotein-complex peptides have been isolated, demonstrating the probable direct covalent interaction of a dentine phosphoprotein with the collagen of the mineralized matrix[640].

The glycosaminoglycans of root dentine from permanent human teeth comprise chondroitin-4-sulfate as the major glycosaminoglycan, and chondroitin-6-sulfate, hyaluronic acid, dermatan sulfate, and a non-sulfated galactosaminoglycan in minor quantities[309].

Osteodentine matrix has a higher concentration of sulfur than dentine matrix. The sulfur detected is presumed to be contained in acid mucopolysaccharides, which were more heavily dristributed in the osteodentine matrix than in dentine matrix[310].

2.2.6 Elastin

For nearly half a century the elastic fibre has been considered a primary site of vas-cular calcification[326, 334–336]. Calcium deposits in the arterial wall commonly take the form of phosphate salts which are intimately intermeshed with the protein of the elastic fibre[327]. Much effort has been directed toward identifying the mecha-nism for the initiation of focal calcification. It has been proposed that the initial step is ionic calcium binding to the uncharged peptide carbonyl oxygens of the elastic fibre protein[328, 329].

Native elastin is an insoluble, highly cross-linked protein. The chemistry, proper-ties and structure of elastin have been frequently reviewed in the past[330–333]. In this context, aspects of calcification processes will be discussed.

It has been shown that vascular mineralization is associated with the elastin fibres[334] of the components on the major arterial vessels. Elastin isolated from human aortas of varying ages undergoes mineralization *in vitro* at a faster rate when obtained from older tissues. This increase in the rate of mineralization may be seen in the reduction of what is called the "lay" period, in the uptake of calcium and phosphate during the *in vitro* calcification of elastin[337].

A lay period in the uptake of calcium and phosphate was previously seen in the mineralization *in vitro* of human aortic tissue[338]. The lay period found with the elastin samples appears in the result of the early stages of the mineralization process, and involves the formation of nuclei for HA crystal growth. Once a microcrystal of HA is present, the rate of calcium and phosphate uptake will increase markedly, because crystal growth and secondary nucleation by the crystal involve a much more rapid incorporation of calcium and phosphate than does nucleus formation. This suggests that the decreased lay period seen in elastin samples from older individuals may be the result of more rapid nuclei formation.

There are two possible explanations for the decreased lay periods and possibly increased nucleation rates in the older tissues:
(1) a change in the amino acid constituents of elastin itself may result in a change of mineralizing properties,
(2) a more firm complexing of elastin with other organic components of the older aortic tissues may cause an increase in the rate of *in vitro* mineralization of the complex.

The latter possibility appears the more likely of the two, especially in view of the results of the enzymatic digestion. These indicate that most of the material isolated from the human aortas was elastin, but some contamination with other substances did occur[338].

Several possibilities concerning the function of amino acids in elastin for mineralizing processes have been discussed. The increase in the polyfunctional anionic amino acids[338, 339] has been suggested to be the result of changes within the already formed elastin molecule or to be due to firm association with a proteinaceous substance rich in these amino acids. Also an increase in desmosine, isodesmosine and lysinonorleucine in older human elastin[340] may change mineralization properties. Several other possibilities for calcium-binding sites in elastin have been suggested. Some include sulfhydryl groups[341–343], carboxyl groups[344–346], threonine hydroxyls[346], and amino groups[347]. More recently, neutral sites in elastin have been proposed for calcium-binding[328]. The sites utilize, particularly from a conformational viewpoint, the most striking feature of the amino acid composition, that is, the high glycine content. Glycines favour the formation of β-turns and associated conformations that are known (from studies on ion-transporting antibiotics) to interact with cations. By analogy with certain antibiotics which are uncharged polypeptides and depsipeptides that bind cations by coordination with neutral acyl oxygens, it is proposed that calcium ion binding also utilizes uncharged coordinating groups, e. g. neutral sites, in the protein matrix. The protein matrix which becomes positively charged by virtue of the bound calcium ions attracts neutralizing phosphate and carbonate ions which then allow further calcium ion binding. The driving force is, therefore, the affinity of calcium ions for the neutral nucleation sites.

The charge neutralizing theory of calcification suggests a fundamental role of organic anions, e. g. sulfated mucopolysaccharides, in regulating bone formation and in retardation of atherosclerosis[328].

Calcium in this case would coordinate with acyl oxygens from the polypeptide backbone of the protein, because of its unique amino acid sequences and potential conformations[348]. Non-polar peptides have been isolated from elastin which consists almost entirely of the three non-polar amino acids glycine, valine, and proline[349]. Portions of porcine tropoelastin have been partially sequenced[350]. Repeating tetra-, penta-, and hexapeptides have been observed. The tetrapeptide contains the sequence —Gly—Gly—L—Val—L—Pro—; the pentapeptide —L—Val—L—Pro—

Gly—L—Val—Gly—, and the hexapeptide —L—Ala—L—Pro—Gly—L—Val—Gly—L—Val—. These sequences are analogous to those found in some of the ion-transporting antibiotics[328]. It has been proposed that these sequences may uniquely contribute to the calcium-binding capacity of elastin, because of their unusually high glycine content. An important conformational aspect of glycine residues is that they can affect the insertion of near right angle turns in a polypeptide backbone[351]. If glycine is followed by an amino acid residue with a bulky side chain, e. g. L—Val, these right angle turns or β-turns become stable conformations[352]. A feature of this type of conformation is its capacity to bind selected ions by coordination of ion with exposed peptide oxygens in the peptide backbone[328, 353, 354].

The cyclododecaptide (Gly—Gly—L—Val—L—Pro)$_3$ interacts with calcium with complete retention of the peptide protons upon titration with $CaCl_2$. N-formyl and O-methyl derivatives of α-elastin act as calcifying matrices[355, 358, 359]. α-elastin and insoluble elastin bind more calcium after presumably a conformational change[356].

With respect to calcification of elastin, the role of neutral binding sites does not preclude the possibility that sulfhydryl groups or chelation sites involving amino acid carboxyl groups represent other nucleation sites[356]. Calcium-binding at neutral sites would result in a positively charged complex, which in turn may direct or modify the association of components comprising the elastic fibre[352].

2.2.7 Egg Shell Constituents

Calcareous egg shells are produced by birds, some reptiles, and gastropods as well as by monotremes. The calcified egg shell protects the content of the egg from mechanical effects during the extrauterine incubation, and it permits an exchange of gases and fluids through pores. Furthermore, egg shells of animals living in arid regions particularly in the tropics and subtropics prevent an excessive dehydration of the egg and of the growing embryo. The relatively thin shell makes the transition of thermal energy possible during incubation period even in low temperature environments (penguins). Moreover, the calcareous egg shell has been regarded as a Ca-reservoir for the construction of the embryonic skeleton[28, 363].

The avian egg shell is morphologically divided into the following sections:
1. Shell membrane (Membrana testacea)
2. Biocrystalline layer
3. Cuticle (Tegmentum)

2.2.7.1 Egg Shell Membrane (Membrana testacea)

The avian egg shell membrane seems to be necessary for the formation of the biocrystalline layer of the egg shell. Under experimental conditions no egg shell is formed in the absence of the membrana testacea (MT)[360]. It is generally accepted, however, that the initial processes of calcification are located in the mammilary layer of the shell[28].

The MT can be divided mechanically into two morphological layers each consisting for fibre networks forming meshes (diameter: 8—10 μm). The meshes can be

83

important for the penetration of microorganisms. The single fibres are constructed of a core and a mantle.

The predominant components of the MT are proteins. In contrast to the usual assumption that these membrane proteins are insoluble proteins, it has been shown that at least 6 % of the total proteins of the inner membrane are water or salt soluble[361]. The extraction of proteins from the MT is extremely difficult with the techniques usually applied in protein chemistry[164, 362]. The egg shell membranes were studied by several authors[364, 365] and some reported that the membranes were of keratin or ovokeratin[366]. The keratinous nature of the proteins has been questioned since a comparison of the known properties of soft and hard keratins with the MT proteins shows distinct differences based on the structure and amino analyses[361] as well as on solubility studies[367]. MT proteins have been regarded as collagen-like proteins because of the presence of hydroxyproline[368] and hydroxylysine[369]. The fibre core has been reported to contain glycoproteins[368]. Also elastin-like proteins have been found in the fibre mantle[376]. The possibility of the attack by a nucleophilic reagent as CN^- on the sulfur atoms of the sterically accessible disulfide bonds of cystine molecules in order to solubilize MT proteins has been studied [371]. The soluble organic material from MT after treatment with CN^- corresponded to $\cong$ 45 % of the intact MT. Structure analysis of the fragments of solubilized MT proteins revealed that about 45 % of the intact MT is constructed of identical repetitive peptides. Since no other fragments could be observed, it has been concluded that those repetitive peptides form the only keratin-like protein in the egg shell membrane. Other proteins of the MT are of different structures. The presence of sulfated glycoproteins in MT has been described[372]. MT contained, when analyzed for sugars, glucose, galactose, mannose, xylose and sialic acid[361]. In addition, hexosamines[365] have been described while uronic acids have not been detected as yet[368]. Lipids have been found too and have been characterized[136, 137] as monoglycerides, cholesterol, diglycerides, free fatty acids, triglycerides, cholesterol esters and PL. The PL fraction included lysolecithine, lecithine, kephaline and sphingomyeline. Also minerals have been detected in MT including P, Ca, K, Na, Mg, Zn, Mn, Fe, Cu, B, and Al[361]. Binding studies with $^{45}CaCl_2$, however, demonstrated that the MT is not directly involved in the initial phases of calcification by binding Ca^{2+}. On the other hand, the possibility cannot be excluded as yet that the regular primary structure of the main MT protein provides positions by which the components of the egg shell formation processes are held in favourable confguration. A more likely possibility is that the network of MT acts as a solid, screen-like supporting structure to keep $CaCO_3$ crystals in appropriate positions[371]. The presence of porphyrins has been presumed[373].

2.2.7.2 Biocrystalline Layer

The biocrystalline layer of avian egg shells is approximately 270–300 μm thick (chicken). It can be divided into morphological units as follows: eisospherites, organic matrix core, zone of tabular aggregates, zone fish bone patterns, external zone, cuticle, and organic matrix. Details of the morphological structure have been reviewed recently[28, 374].

Each layer of the egg shell contains organic material usually referred to as organic matrix. The organic matrix is organized by fibres which differ clearly from those of the MT. Also the types of organic matrix fibres vary from layer to layer as far as diameter, lenght and stainability are concerned.

The complete biocrystalline layer contains ca. 98 % inorganic and about 2 % organic material. The predominant inorganic component is calcium carbonate (98.43 %) as calcite (only traces of aragonite have been reported). Furthermore, magnesium carbonate (0.84 %) and tricalcium phosphate (0.73 %) have been found. The presence of traces of Fe, Cu, J, Mn, Mo, Zn, and other elements has been detected. These described elements are not regularly distributed within the biocrystalline layer, e. g. larger amounts of Mg and P have been observed in the external zones of the shell[375].

The organic substances of the biocrystalline layer include primarily protein-polysaccharide complexes while the contents of water and lipids are low[376]. Polysaccharides composed of glucosamine, galactosamine, mannose and fucose have been detected. Also chondroitin-sulfate-A and dermatan sulfate have been found[368]. Sialic acid is present only in small amounts, which is highest in the mammilary layer and drops gradually to the external zone[377].

The evidence for the presence of the proteinaceous material in the biocrystalline layer is based on amino acid analyses[368]. Systematic studies of the amino acid composition of the biocrystalline layer of egg shells from Galliformes, Gruiformes and Anseriformes and reptiles gave hints to different development within orders of the classes Aves and Reptilia[378-380]. From the data of the amino acid composition of ratite egg shells, affinities could be derived[362].

In egg shells of reptiles, three different types of egg shell protein with a high proline content connected with the presence of cysteine. Another type of reptile egg shell proteins is proline-rich but free of crysteine (first observed in Phelsuma egg shells). A further type of egg shell proteins has first been observed in Tarentola (gekko) egg shells; it is practically cysteine-free, but contains high proportions of aspartic acid, glutamic acid, lysine and serine[380].

It has been concluded from the data of amino acid analyses of the biocrystalline layer of egg shells that the protein or protein derivatives (glycoproteins, proteoglycans etc.) express evolutionary tendencies within the classes Aves and Reptilia. A much better indicator for those developments could be the distribution of proteins or their complexes with other substances. However, since proteins of the biocrystalline layer of the shell are difficult to solubilize without destruction strong efforts were necessary to find means to solve this problem. The first methods for the extraction and fractionation of proteinaceous marcromolecules from the calcareous layer of the egg shell of hens have been described as late as 1975[381, 382]. More than 50 different fractions have been isolated and characterized. This technique included gel filtration, ion exchange chromatography, and isoelectric focussing. When this procedure was applied to the biocrystalline layer of egg shells from birds of different species, species-specific differences could be observed, and even subspecies-specific profiles were obtained. *Gallus bankiva* had a typical protein distribution while *Gallus gallus* again had a characteristic protein profile. Even race-dependent protein distributions have been revealed (Rhode Island Reds, White Leghorns, HNL-hybrids)[363].

A given shell protein profile can be influenced by environmental poisons, e.g. Cd, Pb, Hg, Be[363, 383−386].

In the same context the phenomenon of egg shell reduction has been discussed since thin egg shells are correlated with changed protein profiles of the biocrystalline layer.

Biosynthesis of egg shell proteins has been regarded to be affected by genetic as well as exogenic factors. Animals with a high egg production usually have an egg shell profile lacking one or more fractions, while e. g. *Gallus bankiva* with a low egg production has been demonstrated to have a much more complex protein pattern[363]. But what are the biological functions of the separated protein derivatives extracted from the biocrystalline layer of the egg shell? Using histochemical methods, the presence of carbonic anhydrase in the mammilary layer of the egg shell has been demonstrated[387, 388]. Other authors were not able to corroborate these findings[390−392]. This controversial opinion has led to a reinvestigation of this problem because carbonic anhydrase has been identified in uterine mucosal cells of hens not bound to subcellular structures[390, 393]. High carbonic anhydrase activity in the shell gland of laying hens is correlated with high egg production; but during moult a reduced carbonic anhydrase activity in the egg shell gland and a constant carbonic anhydrase activity in erythrocytes has been observed. It has been assumed that the structure of this enzyme in the uterine cells depends on the action of progesterone[394]. Similar observations have been made studying carbonic anhydrase activity in the endometrium of mammals[395]. A dependence of progesterone action on carbonic anhydrase activity in the mucosal cells of the shell gland of chicken has been found very recently[396]. Obviously, carbonic anhydrase activity in the shell gland of birds is associated with the formation of the biocrystalline layer of the shell since inhibitors of this enzyme, e.g. acetazolamide and other sulfonamides prevent the complete formation of egg shells. In some cases, no shell at all was formed although the Ca level of the blood was normal. However, other mechanisms of action of sulfonamides, e.g. the renal loss of bicarbonates, have also been discussed[397].

Very recently the isolation of still active carbonic anhydrase from the biocrystalline layer of hens' egg shells has been reported. Crude egg shell extracts exhibited no enzyme activity; however, after removal of Ca^{2+} and simple gel filtration of the extracts, carbonic anhydrase activity could be measured. Further purification steps led to the isolation of two isoenzymes. The possibility of inhibition of this enzyme activity by shell components cannot be excluded although recombination of shell components after separation did not inhibit that enzyme. The isolated enzymes has an apparent molecular size of 28,000 daltons. Carbonic anhydrase from egg shells catalyzes the reversible hydration of $CO_2 + HOH \leftrightharpoons H^+ + HCO_3^{1-}$. This probably is the primary action of the enzyme of the shell. Moreover, egg shell carbonic anhydrase catalyzes the hydration of acetaldehyde and pyridine aldehyde. Furthermore, the same enzymes have esterase activity (hydrolysis of *p*-nitrophenyl acetate). Whether the latter activities play a role in the egg shell cannot be judget at the present time.

Carbonic anhydrase activities of the egg shells are inhibited by reaction with *p*-mercuribenzoate by blocking reactive SH-groups of the enzyme. Omission of 2-mercaptoethanol from buffer solutions for extraction and fractionation also resulted in a loss of activity[398-401]. Inhibition by sulfonamides of isolated enzymes from egg shells could be associated with a direct binding of the sulfonamide group to the zinc atoms located at the active sites of the enzymes by substituting for water ligands[402, 403].

DDT inhibits isolated carbonic anhydrase from the biocrystalline layer of egg shells as well.

Although the overall mechanism of shell formation is not fully understood on the molecular basis, it is very likely that carbonic anhydrase is located in the mammilary layer[387, 388]. Since the mammilary layer of the biocrystalline layer of the shell is rich in mucopolysaccharides which accumulate Ca^{2+}, it seems likely that the presence of carbonic anhydrases might have aided calcification either prior to or after the formation of the organic core by localizing a high concentration of carbonate anion derived from either bicarbonate anions or metabolically produced carbon dioxide. Furthermore, the presence of such an enzyme at the mammilary cores may aid absorption of inorganic carbonate by a developing embryo and therefore would also offer an explanation for the observed localized dissolution of the shell at the site of the mammilary cores in hatched egg shells[28, 363, 404-406].

Ca^{2+} also may be made available from the biocrystalline layer of the shell by changing the pH in a microenvironment. Ca^{2+} could be absorbed for the construction of the embryonic skeleton.

Other soluble organic macromolecules from the biocrystalline layer of the egg shell have been demonstrated to bind Ca^{2+}. At least six different Ca-binding macromolecules have been isolated and most of them have been purified to homogeneity[407].

Some of them have been partly chemically characterized and found to be composed of isoprenoid hydrocarbons, carbohydrates and proteins. The ability to bind Ca^{2+} is connected with the protein moiety. Evidence for the affinity of Ca^{2+} to proteins and peptides is derived from the observation that extraction of hydrocarbons does not reduce the Ca-binding capacity. Enzyme fragmentation of the protein moiety resulted in Ca-binding peptides which do not contain sugars or sugar derivatives; however, phosphate has been detected[408]. Amino acid sequence analysis revealed a high proportion of glycine, serine, alanine as well as valine in subfragments of Ca-binding peptides derived from Ca-binding systems. A glycine-rich peptide has even been found in the biocrystalline layer of dinosaur egg shells[409]. In some Ca-binding subfragments of hens' egg shells a high proportion of glutamic and aspartic acids after HCl-hydrolysis has been observed; however, both acidic amino acids could be detected as their amides within the peptide chain. So it is very unlikely that both amino acids play a major role in Ca-binding. Hydrolysis of an isolated Ca-binding subfragment from a Ca-binding complex of the biocrystalline layer yielded mostly glycine[408]. As discussed later, the structure of Ca-binding systems extracted from egg shells are very complex. Other Ca-binding fragments of Ca-binding systems had sequences in which glycine, valine, alanine and serine play an important role in calcium-binding. Ca-binding systems from egg shells also contain carbohydrates, primarily glucose and fucose, as well as an unknown methylated sugar. Glucosamine and mannosamine as well as uronic acids have been identified as constituents of Ca-binding systems. Sialic acid appears in traces only. Sulfate and phosphate have been found in large quantities in intact Ca-binding systems. After treatment with alkaline phosphatase, Ca-binding fragments were obtained which did not contain any hexoses or aminosugars but only traces of uronic acids and phosphate but a relatively high proportion of ester sulfate. Cleavage of Ca-binding systems of egg shells with proteolytic enzymes produced fragments with relatively high amounts of ester sulfate, hexoses, phosphate and uronic acids. However, combined actions of alkaline phosphatase and various proteolytic enzymes such as trypsin, chymotrypsin, thermolysine and pronase yielded small Ca-binding fragments having solely amino acids as building blocks. Through these experiments the "active sites" for nucleation of primary crystals have been identified.

Hydrocarbons are present in Ca-binding systems of egg shells in high proportion (80 %). Extraction of these substances with chloroform-methanol revealed isoprenoid hydrocarbons. Ca-binding is not associated with hydrocarbons since they do not bind Ca^{2+} for the complete Ca-binding capacity remains in the water phase. phase.

When the various Ca-binding systems isolated from the biocrystalline layer of hens' egg shells are allowed to recombine, they form organic crystals (Fig. 1a). The same phenomenon of selforganization is observed with Ca-loaded systems. If HCO^{1-} is then provided (e.g. in the presence of carbonic anhydrase), $CaCO_3$ crystals are formed on top of the organic crystals (Fig. 1b). This system could serve as a model for biomineralization processes during egg shell formation. Selforganization of a water-in-

1a

Fig. 1. Self-aggregation of Ca-binding systems from egg shell matrices. **a)** Highly purified organic Ca-binding systems after recombination; **b)** $CaCO_3$ crystal formation upon aggregated Ca-binding systems after charging with Ca^{2+} and action of carbonic anhydrase

1b

soluble product of originally water-soluble Ca-binding systems, could be compared with the *in vivo* development of a matrix. On the other hand, the *in vitro* formation of a matrix which is charged with calcium is the basis for nucleation sites of the first $CaCO_3$ crystals. Seen in context with the above discussed "active sites" for Ca-binding in Ca-binding systems, the matrix hypothesis can be confirmed on the molecular level. The action of ubiquitously distributed carbonic anhydrase completes the formation of $CaCO_3$ crystals. The same model set up from components of a biological system can be applied to similar components isolated from mollusc shells[410]. Ca-binding systems have been purified from various gastropod and pelecypod shells of recent and fossil species[411, 412].

Ca-binding systems have not only been identified in egg shells and mollusc shells but also in mucosal cells of shell glands as well an in uterine fluid of hens. Ca-binding systems from mucosal cells and uterine fluid resemble each other, but differ in some respects from those of egg shells.

2.2.7.3 Cuticle

The avian egg shell is covered by a tegmentum or "true cuticle" which is morphologically constructed of two layers. The cuticle of freshly layed eggs (chicken) exhibits a reddish fluorescence which is due to porphyrins[373]. The chemical composition of the cuticle has been described to contain proteoglycans with a relatively low carbohydrate moiety. The latter comprises sialic acid, galactosamine, galactose, glucose, mannose and fucose. Uronic acids have not yet been detected nor have acidic mucopolysaccharides been identified[368]. The protein entity is characterized by a particularly high glycine content which is much higher than that of the matrix of the biocrystalline layer and the MT.

2.2.8 Invertebrate Shell Constituents

The organic matrix of mollusc shells consists of soluble and insoluble fractions. The insoluble material is primarily hydrophobic. This is best demonstrated by the affinity of the matrix for lipoidal stains such as Sudan Black, even after thorough lipid extraction. The preponderance of aliphatic amino acids may explain this property[413].

Most biochemical studies of the mollusc shell matrices have been superficial analyses of whole matrix hydrolyzates[414]. These analyses have been reviewed recently[214, 415, 416]. Some studies have demonstrated that the insoluble portion of the matrix consists of more than one macromolecule[389, 417, 418]. Little consideration, however, has been given to the possibility that a portion of the matrix may be soluble in the solution which is used to remove calcium carbonate[414]. A water-soluble fraction of gastropod matrix has been reported[419]. The amino acid composition of a HCl-soluble fraction of etherid shell matrix was reported[420]. Also the molecular conformation of proteins present in the mollusc shells and those of the decalcified films of the nacreous layer were measured for the regions of $4,000-700$ cm^{-1}. The decalcified films were found to consist primarily of proteins. The structure of proteins constituting the decalcified films of the nacreous layer is coiled and has a β-conformation. Possibly, the studies were carried out without any purification of shell proteins[411, 421]. A soluble fraction of mollusc shell matrix has also been identified by ultrastructural studies[422, 423].

The first report on the extraction and at least partial purification of the soluble matrix from *Mercenaria mercenaria* came from CRENSHAW[414]. This author isolated a water-soluble glycoprotein which was evenly distributed through out the shell. This material proved to be homogeneous as determined by gel filtration and electrophoresis. The apparent molecular weight was in the range of 160,000 daltons. Titration of the soluble matrix with $CaCl_2$ in the presence of excess sodium, potassium and magnesium showed that calcium was selectively bound. Urea concentrations above 3 M eliminated the calcium-binding capacity. It has been suggested that the soluble matrix plays a role in the initiation of calcification[414].

A soluble fraction extracted from *Crassostrea virginica* yielded a maximum molecular weight of the order of 1 million daltons. Also from *Crassostrea irredescens*, *Mercenaria mercenaria* and *Nautilus pompilius*, soluble proteins have been isolated and analyzed for amino acid composition[424]. It has been presumed that a significant proportion of the soluble protein of the organic matrix of mollusc shells is composed of a repeating sequence of aspartic acid separated by either glycine or serine. This regularly spaced, negatively charged aspartic acid molecule may function as a template upon which mineralization occurs[424]. However, the amino acid composition of the soluble protein of *Mercenaria mercenaria* does not correspond with that purified by Crenshaw[418, 424]. Unfortunately, it is not clear whether aspartic and glutamic acids are amidated or not.

Fossil glycoproteins of the soluble organic matrix are present in an 80-million year old mollusc shell from the Late Cretaceous Period. Discrete molecular weight components, as determined by gel electrophoresis, are preserved. A particular repeating amino acid sequence $(-Asp-Y-)_n$ found in contemporary mollusc shell proteins was identified in fossil glycoproteins[425].

The calcium-binding protein from *Mercenaria mercenaria* has been shown to be a highly sulfated glycoprotein. Its amino acid composition was very similar to those of mollusc shell matrices except that the soluble matrix had a higher aspartic acid content. All of the dicarboxylic amino acids are present in the protein as their amides. In a number of ways the calcium binding of the soluble glycoprotein very closely resembled calcium-binding properties of a glycoprotein isolated from porcine costal cartilage[414].

A systematic study of the soluble components of the matrix of archeo-, meso-, and neograstropod shells revealed fractionation patterns which are essentially constant within a species. In a given species, the profiles are identical regardless of the habitat, the time of the year when the samples are collected, or the age of the animal. This consistency indicates that these components are a genetic expression for a given species. The profiles vary slightly within a genus, e. g. *Strombus gigas* and *Strombus aureus.* The apparent molecular weights of Ca-binding components from shells of eleven species are not larger than 20,000 daltons. In *Nassa reticulata* several Ca-binding systems exist which can be purified to homogeneity. As discussed in context with Ca-binding substances from birds' egg shells, Ca-binding systems from mollusc shells also show the phenomenon of self-aggregation[410, 411]. Thus, Ca-binding systems from mullusc shells can be aggregated to organic crystals charged with Ca^{2+}, and they react with HCO_3^{1-} forming inorganic $(CaCO_3)$ crystals upon the organic crystal layer.

Ca-binding systems have also been isolated from pelecypods *(Crassostrea gigas)* and fossil oyster shells *(Ostrea carrissima)*[426]. In mollusc shells no carbonic anhydrase has been found. However, polyphenol oxidase activity has been detected in shells of *Crassostrea gigas* and *Strombus aureus*[427]. The enzymes from both species have been highly purified and partly characterized[412, 427]. It has been assumed that polyphenol oxidases in mollusc shells catalyze the formation of polyhydroxylated aromatic amino acids which in turn could form a network as has been discussed for resiline[412, 427] since L-dopa and halogenated tyrosine have been found in the mantle and the periostracum of molluscs[428, 429].

The carbohydrate components of hard-pieces of invertebrates have been reviewed recently[416]. Sulfated polysaccharides have been identified as Ca-binding sites of septal nacre from *Nautilus pompilius*[430]. This indicates that the site is a sulfated polysaccharide (s), and may be a sulfated, calcium-binding glycoprotein similar to the one isolated from *Mercenaria mercenaria*[414].

2.2.9 Coccoliths Constituents

Emiliania huxleyi is a unicellular alga which is surrounded by a number of loose oval discs of calcium carbonate (coccoliths). It has been shown that the $CaCO_3$ of the coccoliths is covered by a thin skin of organic material[431, 432]. It has also been demonstrated that a soluble polysaccharide is closely associated with the coccoliths[433]. The polysaccharide is probably located inside the $CaCO_3$ crystals and therefore termed Soluble Intracrystalline Fraction (SIF). It is also possible that the SIF is protected by $CaCO_3$, but nevertheless is extracrystalline in location. Some of its constituent monosaccharides were identified. Mannose, rhamnose, and xylose are important building blocks while glucose, arabinose and ribose occur in minor quantities. Fucose has not been identified definitively. A number of sugars have not been identified at all. Important is the existence of uronic acids and sulfate. SIF has Ca-binding properties. It binds Ca^{2+} selectively in the presence of Sr^{2+}, Mg^{2+} and Na^{1+}. The presence of two types of Ca^{2+}-binding sites was demonstrated: 0.35 μmole of low affinity sites/mg SIF (dissociation constant 1×10^{-5} M) and approximately 0.6 μmole of high affinity sites/mg SIF (dissociation constant 4×10^{-4} M). Approximately 1 mole of Ca^{2+} is bound per 2 moles of SO_4^{2-}. In this context it should be mentioned that the SO_4^{2-}-containing matrix from *Mercenaria mercenaria* was also found to bind 1 mole of Ca^{2+} per 2 moles of SO_4^{2-} [314]. The same ratio was found for Ca^{2+} binding to a glycoprotein isolated from porcine costal cartilage[434-436].

3 Cellular Aspects of Biomineralization

Our understanding of the mechanisms of biomineralization is superficial[435]. In the past 20 years, most researchers have concentrated upon the extracellular concepts of epitaxy, matrix composition and solubility products while recently the role of cellular organelles and enzymes have attracted more attention. But there are some accepted principles which necessarily carry theoretical implications[445].

3.1 Basic Problems

There has been a general acceptance of three major changes in the understanding of calcium metabolism:

(1) that the calcium ion is pharmacologically one of the most disruptive substances for all normal cell functions;

(2) that intracellular calcium ion concentrations are carefully regulated at levels of 10^{-6} to 10^{-7} moles /l; and

(3) that a surprising variety of cells intracellularly deposit minerals that are rich in calcium salts. Calcium has been regarded as a toxic ion that must be removed from most cells and the occurrence of calcium deposits may therefore represent a form of *detoxification*. Intracellular minerals appear to be of two main types. They are either crystallographically pure deposits containing well oriented crystals or they are of very variable composition occuring in forms that are often "amorphous", concentri-

cally layered and spherical in shape. This latter type occurs in virtually every phylum of animals in a wide variety of tissues[446]. When these granules have been analyzed they were shown to consist mainly of Ca, Mg, PO_4 and CO_3 ions in widely different ratios. More detailed analyses of these granules have shown that they contain small amounts of a wide range of heavy metals. Thus Ag, Al, Cd, Co, Cr, Fe, Mn, Pb, and Zn have been detected in the kidney and the digestive glands of the bivalve *Pecten maximus* and *Chlamys opercularis*. It has been shown that not only the body load of these heavy metals was restricted mainly to these two organs but also demonstrated that both these systems produced intracellular granules with which the metal ions were largely associated[447]. Similar observations showed recently the presence of Mn, Ba, Sr, Fe, Zn, Cu, and Si in the intracellular granules of a variety of insects[448], and Al, B, Cu, Fe, Mn, Si, Sr, and Sn in the granules of various cestodes[449]. It has been found that certain mid-gut cells of the barnacle *(Balanus balanoides)* produced granules of virtually pure zinc phosphate[450]. Studies on *Helix aspersa* have shown that intravascular injections of ^{65}Zn lead to a large incorporation of this isotope into the intracellular granules of the hepatopancreas. It appears likely, therefore, that granule formation may function as a cellular route for the detoxification of heavy metal ions.

First it is worth considering the implications of the suggestion that biomineralization may be a cellular detoxification mechanism for one of the pharmacologically most active cations, e. g. calcium. This is not a new concept[451], but it opens up the possibility that the removal of calcium by precipitating it as a highly insoluble deposit may be energetically more economical than pumping it out of the cells into a supersaturated body fluid.

Second, if biomineralization did evolve from the same system with a detoxification function, then this may throw some light on the process involved. Thus, the mineral deposits that occur are mainly the cationic salts of the major inorganic buffers of cells. The ratios of carbonate and phosphate ions are, however, variable so that it is unlikely that they are the driving forces in the system, and it is in fact easier to imagine a proton extrusion system, operating across a membrane-bound vesicle, as the fundamental process. This would lead to the accumulation of phosphate and carbonate ions within a vesicle, thus producing a general "sink" capable of removing the heavy metal ions from the cytoplasm. On this suggestion, proton extrusion would form the basis of a cellular detoxification system and biomineralization may represent specific forms of this process[452]. An alternative mechanism for removing a variety of heavy metal ions would be to bind them onto organic ligands and to excrete these via secondary lysosomes which eventually release their products into the same system that produces the granules. This could account for the concentric layering of most of the intracellular granules and for the occurrence of various secretions into these vesicles[453]. In this situation one might also expect some selective advantages to accrue from organic ligands which were also capable of being incorporated into the matrix of the mineralized deposit.

These suggestions may go some way toward explaining the occurrence and composition of intracellular granules and it is easy to envisage how such a system could evolve into a mechanism where the deposits became involved in the variety of purposes normally associated with biomineralization[454].

Five basic principles are accepted in biomineralization carrying primarily theoretical implications.

1. Biological calcification will only occur from solutions which exceed the solubility product of the mineral being formed.
2. Calcification induces acidosis.
3. Intracellular Ca^{2+} is about 10^{-6} mole/l.
4. Mineralized structures contain an organic matrix.
5. Mineralized structures are typically highly crystalline.

Perhaps the simplest approach to investigate biomineralization is to inquire wether the process is basically an intracellular or extracellular phenomenon. Clear experimental evidence to favour one view or the other is not in sight. The *extracellular concepts* have traditionally been applied to corals[571-573], molluscs[574, 575], arthropods[576], and vertebrate bone[577]. But in recent years it has been suggested that mineralization is basically an *intracellular event* which is under the closest control until the mineral substances are occluded or 'pinched off' by the cell[579]. Thus when one comes to consider the mechanisms of calcification, the difficulty could not be greater. The implications of these distinctions are enormous but the data relevant to any decision are often very sparse.

Common features of mineralized tissues in the invertebrates are the following characteristics:

1. The inorganic crystals are typically orientated in a precise way if they are part of a skeletal system. The crystals show evidence of a variety of normal growth processes, e. g. spiral and dendritic forms.
2. The crystals are embedded in an organic matrix. This contains glycosaminoglycans[578] typically in chemical association with protein representing a potentially strong calcium-binding complex[580, 581]. The protein is repeatedly thought to have 'crystal orientating' properties[582]. Lipids are often present in these matrices and have been implicated in the process of calcification[590].
3. Two enzymes are almost invariably associated with calcified tissues. The first is carbonic anhydrase, which catalyzes the hydration of carbon dioxide and which is now known to occur in two forms referred to as 'high activity' and 'low activity' isoenzymes[583]. The second is a general phosphatase frequently identified purely by histochemical means and with increased activity in the alkaline range. These alkaline phosphatases also.exist in various isoenzyme forms[584]. The suggested functions of alkaline phosphatases vary from 'synthesizing' matrix material[585, 586] to removing crystal poisons[587].
4. Variable amounts of citric acid are usually found in calcified tissues[588].
5. Calcification occurs from an aqueous medium which is often present in very small amounts forming a microenvironment around the site of mineralization. This fluid contains a variety of organic components separable by electrophoresis[578]. With the electron microscope, the sites of calcification are frequently seen to contain electron opaque granules.
6. The epithelia-lining regions of calcification contain variable amounts of glycogen and large numbers of mitochondria[588].

3.2 Extracellular Sites of Calcification

3.2.1 Calcium and Hydrogencarbonate Pump

When the idea is accepted that ion pumps exist in plasma membranes, then it is very easy to imagine a mechanism of calcification based on such properties. In such a theory, ions are pumped into the fluid surrounding the site of calcification (Fig. 2). This, therefore, becomes supersaturated and mineral deposition occurs. Various pieces of evidence may be quoted to support such a scheme. In its current form it is usual to note that intracellular Ca^{2+} is very low and is maintained in that state by an outwardly directed pump. This calcium pump is powered either by a Ca-ATPase[589] or by energy derived from the Na^{1+} pump[591]. According to current theory, all cells possess this pump and it is therefore easy to assume it being involved in concentrating calcium at certain extracellular sites. Pumping mechanisms consist of a Na^{1+}/H^{1+} and Cl^{1-}/HCO_3^{1-} or OH^{1-} exchange system. It is typically found in the salt-absorbing epithelia of vertebrates, crustacea, and insects.

This theory of calcification, presented in various forms, is perhaps the favourite of most researchers. It is usually proposed in a vague form in which it is not clear which ions are being transported and whether they are moving across epithelial layers or across cell membranes. The basis for the theory is almost *a priori* and there is no specific evidence to support it. A considerable number of attempts have been made to analyse fluids from sites of calcification. The interpretation of these analyses is complicated by having to correct them for ion binding and by difficulties in calculating the ionic concentration of CO_3^{2-} or PO_4^{3-} in buffer mixtures. Few attempts have been made to correct for Donnan equilibrium effects by dialyzing the fluids against an equivalent saline[592], but this is essential if the information is to indicate that pumps have modified the relative ratios of ions. It is highly unlikely that any indication of effective supersaturation will be obtained from such studies since the expected solubility products are largely unknown[593, 594]. In mineralizing systems which form slowly, such as those found in invertebrates, one would expect the fluid adjacent to the minerals to equilibrate rapidly.

Investigations of ion movements across epithelia have, however, only rarely been undertaken[595, 596]. Using the mantle of the lamellibranches, *Anodonta grandis* and *Amblema costata*, it has been shown that there was normally a transepithelial potential difference of 25 to 50 mV with the shell side positive. There was, however, no

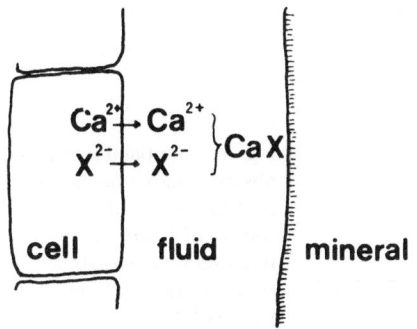

Fig. 2. The ion pump theory. Ions (Ca^{2+} and $X^{2-} \triangleq CO_3^{2-}$ or HPO_4^{2-} are pumped into the fluid surrounding the site of biomineralization. The fluid becomes supersaturated and mineral deposition occurs. (After Simkiss[445])

evidence for the acitive transport of calcium toward the shell and the tissue did, in fact, behave as a calcium electrode with calcium positively diffusing across the external membrane of the epithelial cells. This membrane appeared to be nearly impermeable to intracellular anions so that variations in extra-epithelial calcium levels gave results virtually in keeping with NERNST's equations

$$\psi = \frac{RT}{zF} \ln \frac{Ca_i}{Ca_e}$$

The results obtained are of fundamental importance because they show no evidence of a calcium pump, no evidence of facilitated anion movements, but a zero potential at 12–16 meq/l Ca^{2+} [595]. More recent work investigating profiles of the potential difference across these tissues have cast some doubt on this interpretation[596]. Calcium electrode effects are not restricted to the mantle epithelium of molluscs[597] (Fig. 3). At a low concentration of calcium there may be an active uptake of this cation from the water by snails[598] and crustacea[599].

3.2.2 Counterion Movement

A number of suggestions have been made that calcium may be transported because it is coupled to the movement of other ions or because it moves passively down an electrochemical gradient established by the movement of some other ions[600]. Thus, a sodium-induced potential has been found which was sufficient to account for the passive movement of calcium into the shell gland of the domestic fowl during egg shell formation. In the mollusc, the shell side of the mantle is normally positive relative to blood and a potential of this type would, of course, tend to move calcium away from the extrapallial fluid. A potential of this orientation could be produced by the movement of an anion into the animal (mollusc) and the low chloride concentration of the extrapallial fluid could be accounted for on this basis.

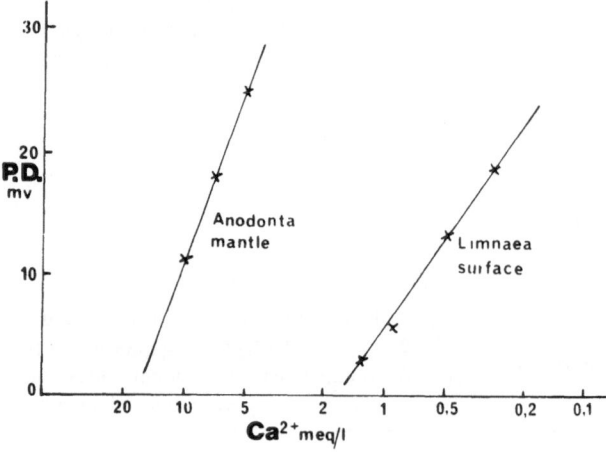

Fig. 3. Potential difference observed across the mantle of *Anodonta* or the body surface of *Limnea* as influenced by external Ca^{2+} concentrations. (After Simkiss[444])

3.2.3 Proton Removal

Biomineralization is basically regarded to be an acid-base problem[601]. In addition, if there is no clear evidence for either calcium or bicarbonate pumps at sites of calcification, the alternatives should certainly be considered. This theory of calcification suggests, therefore, that the basic energy source is a removal of protons, i. e.

$$Ca^{2+} + HCO_3{}^{1-} \rightarrow CaCO_3 + H^{1+} \tag{1}$$

At pH 7.0 the hydrogen ion concentration is 10^{-7} mole/l. A proton-"catching" system using a metabolically driven pump might therefore be considered to be unlikely when compared with sodium, for example, where the intracellular concentration would be several thousand times greater than this. The secretion of hydroxyl ions might therefore be considered a more likely basis for proton removal, i. e.

$$Ca^{2+} + HCO_3^- + OH^- \rightarrow CaCO_3 + H_2O \tag{2}$$

A variation of this reaction is perhaps more revealing, i. e.

$$Ca^{2+} + CO_2 + 2OH^{1-} \rightarrow CaCO_3 + H_2O \tag{3}$$

There is a good deal of support for such an origin of the carbonate ion. Certainly, in invertebrates there are conclusive kinetic data based on radioisotope studies to show that, although the blood supplies Ca^{2+} for egg shell formation, the plasma bicarbonate plays no direct role[602].

It may not be immediately apparent from Eq. (1) to (3) exactly what effect hydroxyl ion secretion has upon sites of calcification. The scheme can be written as follows:

$$CO_2 + OH^{1-} \rightleftharpoons HCO_3{}^{1-} \tag{4}$$

$$Ca^{2+} + HCO_3{}^{1-} \rightarrow CaCO_3 + H^{1+} \tag{5}$$

$$H^{1+} + OH^{1-} \rightleftharpoons H_2O \tag{6}$$

The reaction of Eq. (4) would, of course, be capable of being catalyzed by carbonic anhydrase. Equation (5) represents the formation of mineral and the release of protons, whereas Eq. (6) indicates the removal of protons. Thus, although hydroxyl ion secretion may be the motivation force, the immediate result may be a slight fall in pH at the site of calcification. This is in keeping with the data, pointing out that the extrapallial fluid of molluscs is slightly more acidic than the blood.

A variation of the proton removal hypothesis of shell formation has been proposed[603, 604]. It has been shown that in molluscs and birds the sites of shell formation contain high concentrations of ammonia and they suggest that this may be an initiating phenomenon for calcification, i. e.

$$Ca^{2+} + HCO_3{}^{1-} + NH_3 \rightarrow CaCO_3 + NH_4{}^{1+} \tag{7}$$

There seem to be two problems: NH_3 is non-charged and mobile and easily crosses plasma membranes. When charged, however, in the form of NH_4^{1+}, it penetrates cells with difficulty and tends to accumulate in biological systems. As a proton remover, therefore, ammonia secretion is a poor mechanism. It may temporarily trap the protons, and therefore facilitates the reaction as shown in Eq. (7), but the products tend to remain at the site of cacification.

One is left therefore with the question, does the accumulation of ammonia induce mineralization or vice versa[583].

3.2.4 Matrix Concepts

The matrix is thought to be important in binding various ions at anionic or cationic sites. The organic matrix and its role in biomineralization constitutes the organic matrix concept, which in a broad sense visualizes the organic component as a precursor template of sheets, fibres, or compartments onto or into which nucleation of the inorganic phase takes place and onto or into which the crystals grow and ultimately fit together to form the final mineralized tissue. Although the mineral composition of most calcified material is reasonably well known, the size, shape and orientation of the constituent crystals is a matter of considerable confusion. Several reasons for this deficiency exist but in part the problem is a technical one related to the microscopic nature of most shell structures. The organic matrix concept is the basic idea underlying current theories of biomineralization. It is thus recognized that mineralized tissues are biphasic and that the organic phase is a precursor to the inorganic phase. In this way it is conceived that the organic matrix acts in some way to control the mineralogy and crystallography of the inorganic phase. But at this point divergence of opinion exists, resulting in two major schools of thought: the *epitaxy* school and the *compartment* school. In one, the emphasis is placed on the process of epitaxy as a controlling force while in the other the formation of ordered organic matrix compartments is viewed as the important factor. Both consider the *matrix* to be a distinct structural entity.

The *epitaxy hypothesis* assumes that one substance can take place on the surface of another when the two substances have some crystallographic parameters in common. The stereochemistry of the substrate phase dictates the orientation of the overgrowth phase. When extended to biomineralization, the idea is that the organic matrix takes on the role of substrate or template, thus acting to direct the nucleation and growth of the mineral. The apparent simplicity of this process makes it most appealing and it has thus gained widespread acceptance. The alternative *compartment hypothesis* places little or no emphasis on epitaxy, proposing instead a crystal growth which takes place within ordered organic compartments independent of reactive templates. Essentially, therefore, the ordering of the organic matrix compartments have been thought to result in ordered crystals which grow in the space provided. This theory does nothing to explain the ordering of the crystals, but only shifts the question to the mechanisms which cause the ordering of the organic compartments[437].

There has been a major controversy about both hypotheses. An interaction comparable with epitactic processes has been assumed by several authors[28, 438–442].

Actually, there is no experimental evidence backing the compartment hypothesis, at least not in mollusc shells[443, 444].

There are, however, two theories of calcification which do not involve these concepts and they will briefly be mentioned since they are relevant to the basic problems of ion movement across epithelia.

The first and most widely applied concept is the *semiconductor hypothesis*[607, 608]. It is perhaps most easily understood in relation to the calcification of arthropod exoskeletons. In the shore crab, *Carcinus maenas*, it is suggested that ions are continually diffusing out of the animal across the cuticle. Since the ions move at different rates, a potential is set up with the outer surface positive. The exoskeleton acts as a semiconductor so that there is an outward flow of electrons leaving the inner surface rich in protons and the outer surface rich in hydroxyl ions. Since an increase in pH at the outer surface favours the formation of carbonate ions, the fluid in the outer part of the cuticle becomes supersaturated with calcium carbonate and crystals are deposited. The theory is ingenious since it only relies upon physiochemical phenomena. A great deal of reliance is placed on the diffusion of sodium chloride out of the animal and it is not clear how an osmoconformer such as *Maia* would calcify its skeleton. If the flow of sodium and chloride ions was reversed, the exoskeleton would, of course, decalcify[607]. The semiconductor properties of the exoskeleton are mainly due to quinone tanned proteins and this has enabled the theories to be applied to other situations where these compounds are found. On obvious situation is, of course, the periostracum of the molluscan shell. Using *Mytilus edulis*, it has been suggested[608] that in this case the periostracum is polarized so as to be acidic outside and alkaline inside, i. e., the opposite polarity to crustacea (Fig. 4). In order to explain calcification in the shell, it is now necessary to have salt flow inward and it is suggested that this is achieved by mechanical suction forces developed by the mantle. The theory has been criticized[575] since it provides no explanation for calcification in shells lacking a periostracum for shell regeneration, for shell formation in terrestrial snails, nor for pearl formation in the pearl sac. It has been also pointed out that it provides no explanation as to why the crustacean exoskeleton is not calcified before moult and why, in fact, the observed rate of calcification is correlated with general body uptake[609]. The streaming potentials have also been suggested to form a basis for bone formation[605].

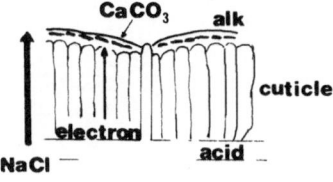

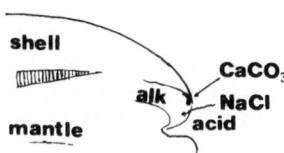

Fig. 4. Simplified version of Digby's semiconductor theory of biomineralization. In the arthropod (top) ions are continually diffusing out of the animal across the cuticle at different rates setting up a potential with the outer surface positive. This causes a flow of electrons leaving the inner surface rich in proteins and the outer surface with hydroxyl ions. The alkaline outer surface favors $CaCO_3$ formation. In molluscs (bottom) muscular movements cause salt flow through the periostracum followed by an alkaline reaction on the inside inducing $CaCO_3$ deposition. (After Simkiss[445])

A somewhat similar theory[610] might also be mentioned here. This theory depends upon an electrical potential difference existing across an organic matrix which acts as a fixed ion exchanger. If this occurs, calcium migrates toward the cathode faster than sodium, which therefore accumulates on the fixed charges. The electrophoretic velocity of chloride is greater than that of phosphate. Thus, both calcium and phosphate accumulate at the cathode while sodium and chloride ions occur at the anode. The result might be a separation of ions favouring the deposition of calcium salts at the cathode.

3.2.5 Ion Concentration Systems

In recent years, great advances have been made in our understanding of fluid movements across epithelia following the standing gradient osmotic flow "Theory of DIAMOND"[611, 612]. The ultrastructure of cells showing this type of fluid movement has prominent lateral intercellular spaces which are open at the serosal surface, but closed by a tight junction on the lumen side. Salt is actively pumped into the lateral spaces which therefore become hyperosmotic. Water passively follows resulting in the formation of an almost isotonic fluid which passes on to the serosal surface. The exact concentration of this fluid will depend upon the dimensions of the spaces and other cell characteristics.

It will be apparent that if "normal extracellular fluids" were subjected to an isotonic resorption of sodium and chloride ions by the process, the net effect would be to concentrate other ions and precipitate minerals. This suggestion was made[613] to explain one of the methods of forming deposits in the calciferious glands of earthworms. It was proposed that the posterior glands received blood directly from the intestine. Fluid was formed in these glands by a process of filtration and saline was then resorbed by the epithelial cells. This resulted in the formation of calcareous deposits (Fig. 5).

Electron micrographs of these glands show structures which would be in keeping with this interpretation[614]. Arising from the basal lamina are numerous membranous infoldings. The lateral membranes are closed by tight junctions at the lumenal surface but freely open into to blood sinuses on the serosal surface. The lumenal surface contains many microvilli (Fig. 6).

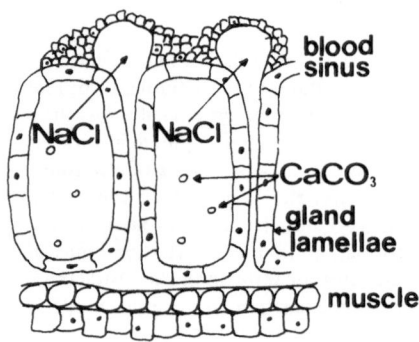

Fig. 5. Ion-Concentrating Systems. The calciferous glands of the earth worm contain extracellular fluid from which there is an isotonic resorption of NaCl. This results in the formation of $CaCO_3$ deposits. (After Simkiss[445])

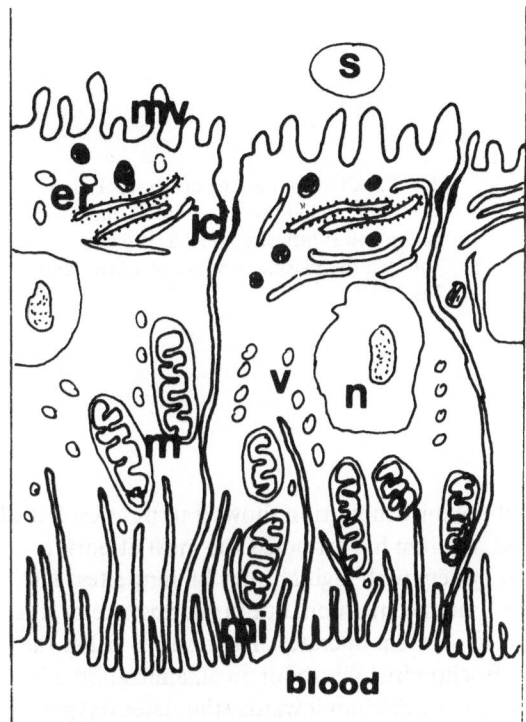

Fig. 6. Ion-Concentrating Systems. The microvilli at calciferous mucosal gland cells of the earthworm resorb ions from the lumen which are moved into the intercellular spaces behind the junctional complex (Jc) followed by an isotonic secretion into the blood. MI = Membrane investigations; V = Vacuoles; N = Nucleus; M = Mitochondria; E. R. = Endoplasmic reticulum; S = Spherolith. (After Simkiss[445])

3.3 Intracellular Sites of Calcification

A variety of theories have been proposed to explain the association of carbonic anhydrase and alkaline phosphatase with sites of cacification and they are basically of two types. Thus, the enzymes are thought to be responsible for producing CO_3^{2-} or PO_4^{3-} ions at the sites of mineralization or of releasing calcium ions. These are all extracellular concepts, but there are also some theories that might be applied to intracellular phenomena. Perhaps one of the more interesting of these is a suggestion[615] involving the chelation of calcium by ATP and its subsequent hydrolysis to produce HA. This theory may be modified following the observation that alkaline phosphatase and Ca^{2+}-ATPase cannot be separated[616].

In this theory, calcium enters the cell down the enormous electrochemical gradient that normally exists between the extracellular and intracellular fluid compartments. In doing so, it actually reverses the calcium pump and synthesizes ATP in a similar way as the reversal of the sarcoplasmic calcium pumps[617] (Fig. 7). There is however, no evidence to support this theory and it has a number of unlikely features. It stresses the need for relevant data rather than circumstantial evidence, and this is particularly necessary in considering intracellular theories.

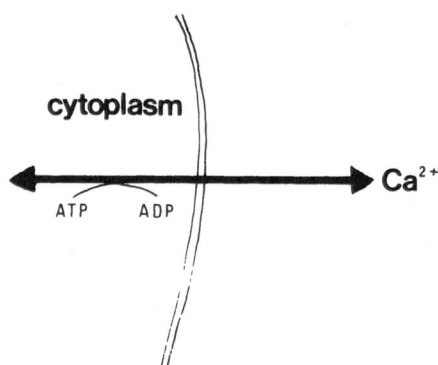

Fig. 7. Calcium moving down its electro-chemical gradient reverses the normal calcium pump and synthesizes ATP as it passes through the plasma membrane. (After Simkiss[445])

3.3.1 Mitochondrial Calcification

It has been known for many years that the mitochondrion shows a respiration-linked transport of a number of ions. Of these, calcium has attracted the most attention since it depends on a specific transport system with high-affinity binding sites. The uptake of calcium usually also involves a permeant anion, but in the absence of this, protons are ejected as the electron transfer system operates. The result is either the accumulation of calcium salts in the mitochondrial matrix or an alkalinization of the interior of the mitochondrion. The transfer of calcium inwards stimulates oxygen utilization but provides an alternative to the oxidative phosphorylation of ADP[618].

The transport of calcium into the mitochondrion can lower external Ca^{2+} to levels of 1 to 0.1 μmole/l. This has, therefore, been interpreted as a basic mechanism in maintaining intracellular calcium at these levels. Only about 3 % of the calcium which passively diffuses into the cell is expelled by a calcium pump into the plasma membrane, whereas the remaining 97 % is sequestered into the mitochondria. This occurs because both processes have similar rate constants, but the total mitochondrial surface is some 30 times larger than that of the plasma membrane. This argument presupposes, however, that the calcium which enters the cell is equally available to both sets of membranes.

It will be apparent that in suitable situations mitochondria will accumulate calcium salts and eventually precipitate them as inorganic granules. Such structures have been observed in electron micrographs of calcifying cartilage[619] and have led to the suggestion that mitochondria may provide the basic mechanism for calcification[620]. The basic systems of calcification involve concentrating calcium and phosphate ions within the organelle[620]. However, both ions are present at much lower concentrations in the cytoplasm than they are in the extracellular fluids. They must, therefore, be "pumped" into the inner compartment of the mitochondrion by specific carriers. When the solubility of the mineral is exceeded, and this may be helped by the alkalization of the inner part of the mitochondria, then the mineral is precipitated as minute electron opaque granules. It is suggested that these are similar to the amorphous phase of bone mineral, being perhaps 20—30 Å in diameter. The formation and dissolution of these granules is presumably reversible since calcium can be made to leave mitochondria by appropriate stimuli. In fact, it is suggested[621]

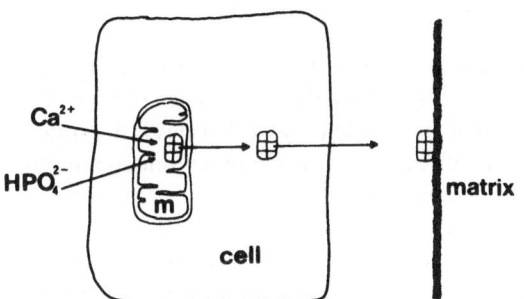

Fig. 8. Lehninger's mitochondrial theory of calcification. (After Simkiss[445])

that these are the means by which mitochondrial granules are transported out of the cell and into the extracellular sites of mineralization. In LEHNINGER's theory, the "micropackets" of mineral either pass directly through the mitochondrial membranes or some form of reversed pinocytosis carries them as membrane-bound particles to the outside of the cell (Fig. 8).

The mechanism by which the mineral leaves the mitochondrion is only one of the problems of this theory. The mineral in the mitochondrion exists in association with the fluid contents. Thus, unless this water is in some structural form with abnormal solubilities, the mineral must be saturating the fluid, and solubility products apply. It follows that the mitochondrial calcium and phosphate concentrations must be similar to those of the extracellular fluids, i. e. calcium must be concentrated thousandfold to overcome the low intracellular values.

Clearly, it is possible to form mineral deposits in the mitochondria but the energy cost is very high. Furthermore, since the fluid inside the mitochondrion is 1000 times more concentrated than the cytoplasmic level, the mineral will continually be tending to dissolve and pass out into the cytoplasm. The low efflux rate constant indicates that leakage is slow, but it represents a nonsteady state and a continual energy drain upon the cell. The contrary argument would be that cells must maintain a low intracellular Ca^{2+}. The mitochondria present one of the possibilities of achieving this concentration. In order for them not to concentrate calcium ions against an ever increasing gradient, it is converted into a mineral form. Having achieved that, mineralization becomes a useful side product.

A third problem with the mitochondrial theory of biomineralization is that many mineralized tissues contain carbonate rather than phosphate. Since bicarbonate ions do not pass across mitochondrial membranes with any ease, it has now been shown that in phosphate-free buffers, calcium will enter mitochondria if dissolved carbon dioxide is available. It appears that some mitochondria possess carbonic anhydrase activity on the inner membrane or in the mitochondrial matrix and are thus able to synthesize bicarbonate within the organelle. In such cases, inhibitors of carbonic anhydrase block the accumulation of calcium and carbonate ions[622] since crystals of calcite have been identified in the mitochondria of earthworms' calciferous glands[623]. These cells freqently showed spherical granules in the cytoplasm and lumen of the glands during phases of mineral secretion and it was suggested that they were aspects of cellular breakdown which occurred at these times.

3.3.2 Calcification in Golgi Apparatus and Endoplasmic Reticulum

This is the most commonly described form of intracellular calcification. Typically, it arises in a smooth membrane which may be formed by the coalescence of several Golgi vesicles. An organic matrix of protein and acidic polysaccharides is deposited within this structure. Frequently, saccules of the endoplasmic reticulum communicate with the vesicle and mineralization occurs within the matrix. Examples occur in most phyla, e. g. the insect midgut[624] and cestode parenchyma[625]. It may be important to recognize two types of crystal deposits from this system. One type is typically a concentrically layered structure containing large amounts of calcium, magnesium, phosphate and carbonate ions; the other is crystallographically more pure and is frequently shaped as in coccoliths[626].

This general theory is sometimes made more precise by considering that the Golgi body is involved in producing the matrix material while the endoplasmic reticulum transfers calcium to the developing vesicle. The endoplasmic reticulum has been studied most intensively in muscle where its ability to transport calcium into vesicles of the sarcoplasmic reticulum is well known. There is, however, some doubt as to how this ability is developed in non-contractile cells[627].

The hypothesis that the endoplasmic reticulum may be an important site for calcification has the advantage over the mitochondrial one in that the movement of vesicles and the opening of the endoplasmic reticulum to the outside of the cell have been observed on a number of occasions. There seems to be little doubt that deposits do leave cells in this way. Gorgonian coral growth has been described as involving a scleroblast attending to a spicule and causing it to grow by means of vesicles containing matrix and minerals[628].

3.3.3 Extracellular Vesicles

Extracellular vesicles or matrix vesicles have not been identified in any invertebrates[445]. In bone formation, it has been suggested that they provide a basis for

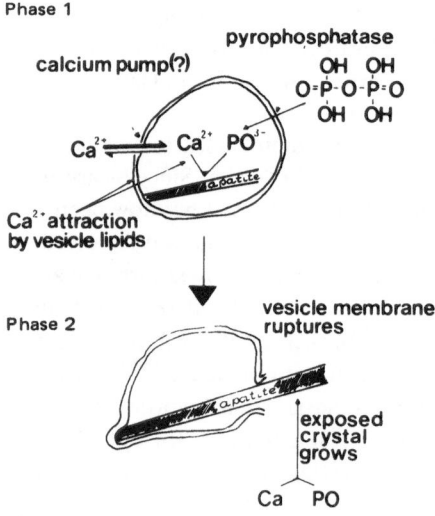

Fig. 9. A generalized form of Anderson's "membrane vesicle" theory of biomineralization. The vesicles break away from the cells adjacent to the site of calcification.
Phase 1: Ions are transported inward and crystal growth is induced within the vesicle. The intravesicular ionic product is raised resulting in the initial deposition of calcium phosphates as HA (and probably also as ACP). Eventually the vesicle membrane ruptures releasing the mineral.
Phase 2: Preformed HA crystals are exposed to the extravesicular environment enabling further crystal growth. (After Anderson[458])

IRVING's association of lipids with sites of calcification. On that basis, it is interesting to note the confirmation of lipids in mollusc shells[629]. Another site where extracellular vesicles might exist is in the crustacean cuticle[630]. The extrusion of granules from the epidermis has been described.

The identification of calcifying membranous extracellular vesicles within the matrix of *cartilage, bone* and *dentine* by electron microscopy[455-457] is one of the most important developments in biomineralization in recent years.

In these tissues the distribution of vesicles correspond closely to the patterns of matrix mineralization. Furthermore, it has been suggested that crystals of HA are deposited within the vesicles. Subsequently apatite is deposited within the vesicles and upon their surfaces to produce typical modular clusters of mineral. Such morphological observations strongly implicate the matrix vesicles in the formation of apatite crystals (Fig. 9). Once the first crystals are formed, they mineralize further by epitactic crystal growth[458].

3.3.4 Fine Structure of Matrix Vesicles

The one structural feature common to cell matrix vesicles is the enveloping membrane which appears trilaminar under the electron microscope. The term "vesicle"

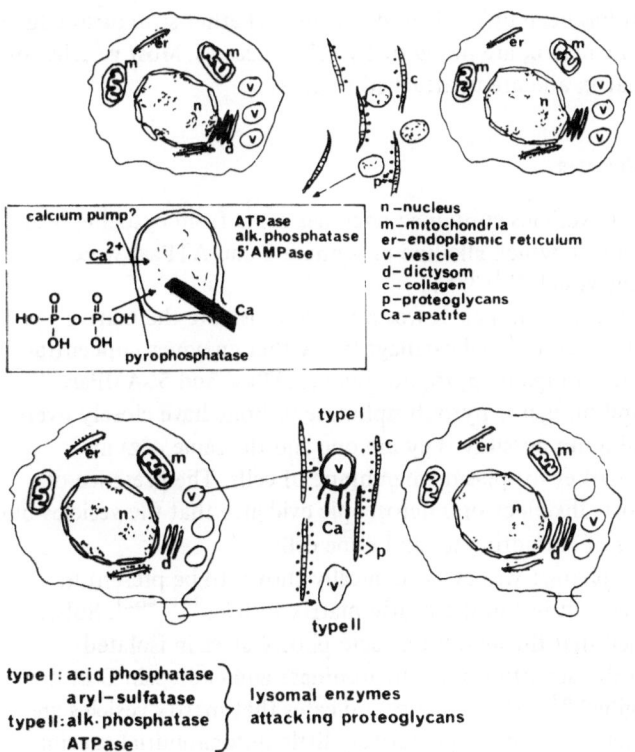

Fig. 10. Schematic presentation of the theories of mineralization in matrix vesicles. a) Theory of Anderson[458]; b) Theory of Thyberg[642]. (After Höhling et al.[641])

was suggested. The presence of the membrane indicates that the vesicles must have come from cells rather than being formed *de novo* in the matrix. The vesicle membrane probably should be considered as a cytoplasmic organelle since cellular membranes often contain enzymes capable of metabolic acitivity, and the membranes of matrix vesicles are rich in enzymes which might function in calcification. It is also likely that the vesicle membranes divide the internal and external environments and enable the sequestration and storage of calcium and phosphate during initial formation of apatite crystals (Fig. 10a, Fig. 10b). Thus, the matrix vesicles can be regarded as a kind of extracellular organelle with enzymes which favour the local development of apatite. The size of matrix vesicles varies considerably, but most reports state the average diameter as about 100–200 nm.

The relation of vesicles to chondrocytes has been studied by electron microscopy[455, 456, 460, 461], and it has been demonstrated that most vesicles are isolated in the matrix and are not connected to cartilage cells[455, 456]. However, if matrix vesicles originate in chondrocytes, then there must be a storage during which the vesicles emerge from the cells. It has been shown that vesicles are located very close to the lateral edges of flattened chondrocytes in the proliferative zone of the epiphyseal plate[456]. It has been suggested that the vesicles arise by budding from cells in the upper epiphyseal plate. After separation, the vesicles are apparently entrapped in that part of the cartilage matrix which is going to calcify[458].

Bone cells, unlike cartilage cells, characteristically possess many long cytoplasmic processes, and so it has been difficult to determine whether structures which appear as vesicles in the bone matrix are not actually cell processes. Most vesicles are isolated within the bone matrix and are unattached to cells.

3.3.5 Enzymes in Matrix Vesicles

The first evidence that matrix vesicles contained enzymes came from cytochemical studies on epiphyseal cartilage in which alkaline phosphatase and ATPase were discovered on the surface of the vesicle[461, 462].

With alkaline phosphatase as a marker, it was possible to isolate the matrix vesicles from other parts of the epiphyseal cartilage[463]. Other enzymes concentrated within the vesicles include inorganic pyrophosphate, ATPase and 5'-AMPase[463]. The alkaline phosphatase and inorganic pyrophosphatase of bone have closely overlapping specificities[464] and quite possibly they are one and the same enzyme. 5'-AMPase is regarded as a marker for plasma membrane of cells. The presence of 5'-AMPase in vesicles supports the electron microscopic evidence that the vesicles bud from the plasma membranes of chondrocytes and bone cells.

The lysosomal acid phosphatase was cytochemically shown to be present in dense bodies of chondrocytes but not in the nearly matrix vesicles[461, 462]. Subsequent studies have confirmed that the amount of acid phosphatase in isolated vesicles is low and also that the activities of β-glucuronidase and cathepsin D in the isolated vesicles were negligible[463]. The evidence indicates that matrix vesicles are not lysosomal. Isolated vesicles contain comparatively little mitochondrial succinic dehydrogenase, suggesting that the matrix vesicles and mitochondria were not identical[458].

3.3.6 Lipids in Matrix Vesicles

The vesicles were shown to be rich in lipids with at laest 1.5 times more total lipid, phospholipid, cholesterol and glycolipid per unit protein than in whole chondrocytes. The ratio of cholesterol to PL was 1.7 times greater in vesicles than in chondrocytes and vesicles which contained twice the cellular amount of phosphatidyl serine and sphingomyelin[102, 465] and which are depleted in phosphatidyl choline. It is of note that the principal PL in vesicles was phosphatidyl serine which has been shown to have a strong affinity for calcium ions, especially in the presence of phosphate[96].

3.3.7 Electrolytes in Matrix Vesicles

Matrix vesicles are markedly rich in Ca compared to epiphyseal chondrocytes, even in the proliferation zone. The Ca and inorganic phosphate concentrations are greater than can occur together in solution. X-ray diffraction of the freezedried vesicles failed to detect any apatite phase[467], while membrane-enclosed vesicle-like structures were found to be filled with amorphous electron-dense material[468]. However, it has been suggested that HA is formed within matrix vesicles[459].

Matrix vesicles isolated from the zone of hypertrophy are much richer in Ca and inorganic PO_4 than those obtained from the proliferating zone. This indicates that not only Ca and P_i concentrations occur during vesicle formation, but that it continues extracellularly after vesicle elaboration into the matrix. Ca/P_i molar ratios of the matrix vesicles from the proliferating and hypertrophic layers are strikingly higher than those of the isolated cells and also higher and more variable than those of the later mineral deposits. This suggests a primary Ca-binding mechanism in matrix vesicle mineralization in which varying amounts of P_i may be concentrated whithin the vesicles by virtue of a Ca–P_i–PL interaction and possibly by active calcium transport across the vesicle membrane[458]. However, an ATP-driven Ca-pump in the matrix vesicles has not been found as yet[469] and vesicles from epiphyseal chick cartilage do not actively transport Ca^{2+}[470] through a membrane-bound ATPase.

3.4 Types of Calcification

3.4.1 Intracellular

It is no longer acceptable to consider the occurrence of mineralized deposits as evidence of calcification. Much of the confusion on the literature exists because of the tendency to associate mineral deposits with the phenomenon of calcification. Once this confusion is removed, the possible types of calcification become much clearer and they can be separated from problems of recrystallization.

This distinction can now be exploited by considering the energy changes associated with calcification. In drawing up these schemes, a major problem is to assign values to the permeabilities between the various compartments. In pursuing this concept therefore, we will oversimplify the situation by only considering the calcium ion and by attributing arbitrary energy values. Some corrections will have to be made for this after the examples have been considered.

The Shapiro and Greenspan theory involves three or possibly four energy hurdles[621]. The first of these is common to all cells, since it is the energy necessary to pump all calcium out of the cell. A second hurdle exists in pumping calcium out of the cytoplasm (at 10^{-6} mole/l) into the mitochondrion where it must reach solubility product values (say 10^{-3} mole/l). The ions thus accumulated are then released into the cytoplasm and pumped out again across the plasma membrane to a level equal to that of the extracellular fluids (hurdle 3) or actually to supersaturation levels (hurdle 4). It may be that one of these hurdles can be omitted in an overall view. Since hurdle 1 is common to all cells, one could simply say that hurdles 2 or 3 were replacing it. One should, however, also note that this static picture ignores leakage from the organelles. Hurdle 2 could be a major problem if its surface area is 30 times that of the plasma membrane and if there is an appreciable back diffusion out of the mitochondria. The theory of Lehninger[620] attempts to overcome some of these energy hurdles by allowing exit from the mitochondria via vesicles. Some evidence for this theory comes from the work on the shell gland of the fowl[631]. These data indicate a movement of calcium from mitochondria to the endoplasmic reticulum during calcification. However, under normal circumstances it is very rare to find granules within the mitochondria of this tissue.

In experiments on the regeneration of the shell in *Helix aspersa*[634] there have also occasionally been found slight increases in mitochondrial and microsomal ^{45}Ca, but in their preparation relatively little ^{45}Ca exists in these fractions at any one time[632]. Furthermore, these data could be reinterpreted as simply indicating that, during mineralization, cells are exposed to a slight calcium shock which stimulates the intracellular mechanisms of calcium homeostasis. Therefore, the only direct evidence, to support the energetically expensive theory are the observations on earthworm calciferous glands[623].

The endoplasmic reticulum theory has much lower energy demands. It has been conjectured that it can produce two types of deposits which different solubilities. The "low solubility product" is a skeletal form. It has well oriented crystals, precise shape, and relatively high purity. Energetically, it is very easy to form. The endoplasmic reticulum has to do no more than to replace the hurdle which is common to all cells. In fact, it may even be energetically economical in that pure crystals may form at a Ca^{2+} concentration which is lower than the external medium so that hurdle 1 is actually diminished. Such a situation may exist in coccolithophorids[626] or corals[628]. The "high solubility product" form could be mineral with many "contaminating" ions. Typical for this type are the digestive gland granules of many molluscs. These were shown to have a $Ca : Mg : PO_4 : CO_3$ molar ratio of $1 : 0.6 : 1 : 0.1$[594]. Similar examples exist in cestodes[633] and many arthropods[634]. They frequently appear to be composed of concentric layers of mineral and, because of the influence of these contaminants, it may be interesting to consider that they represent minerals with a solubility product close to those of the extracellular fluids of most metazoa. If crystallographically "pure minerals" were formed in such fluids, massive recrystallization might occur[632]). These crystals can, by being somewhat impure, exist in the body fluids and thus serve for a large number of functions (e.g. acid base balance, calcium transport, etc.).

3.4.2 Extracellular

The third type of calcification is extracellular. In its simplest form it occurs through the cellular pumping of ions so that eventually the solubility product is exceeded on the calcification site. There is, therefore, only one energy hurdle which need not exceed the normal demands of all cells if the site of crystallization does not contain crystal poisons. The energy demand at hurdle 1 will be reduced if there is epitaxy via an organic matrix. The same energy-saving concept can, of course, be applied to the intracellular theories but it is included here since this is the traditional site. A variation of the extracellular theory is the vesicle concept. The membrane-bound vesicle hypothesis is a very attractive concept since it has all of the intracellular theories but without the problem of existing in a "low calcium" environment. It could, therefore, provide a system with negligible energy demands for the initiating crystal formation.

The final type of calcification mechanism involves some form of extracellular transport system which avoids all the transcellular energy barriers. As a concept, it is extremely valuable but it needs redefining of its relation to the ions under consideration and the motive forces for their movement.

Throughout this review of biomineralization, one outstanding phenomenon has been consistently evaded. This is the problem of apparent solubility products and supersaturation. It is obvious, however, that almost by definition this is the one phenomenon that directly relates to the whole problem, but it is consistently evaded because there is no simple theory to account for the observations. In order to highlight the importance of this phenomenon, therefore, an attempt will be made to relate extracellular theories to this concept.

It appears to be the case that most animals maintain the concentration of "mineral ions" at constant levels in their extracellular fluids. Perturbations with various forms of acidosis usually result in the animal re-establishing an equilibrium between its body fluids and the apparent solubility product of some mineral. Two important conclusions follow from this. First, it provides a theoretical basis for defining calcification. When there is a change of phase in the total extracellular fluids (i. e., mineralization occurs) then the fluids re-equilibrate to make good the ions which have been lost as minerals.

$$Ca^{2+} + CO_3^{2-} \rightarrow CaCO_3 \text{ (mineral)}$$

$$HCO_3^- \rightleftharpoons CO_3^{2-} + H^+$$
i. e., protons are released at the site of calcification.

If, however, mineral at one site forms at the expense of a solution of mineral elsewhere, there will be not net loss or release of protons since the fluid remains with the same ionic composition. This would not be the case without the concept of continual saturation, but accepting this concept it facilitates the distinction between calcification (i. e., *increasing the total mass of calcified material by displacing protons)* and mineralization (i. e., *the deposition of crystals without any change in the total mineral content of the system).*

The second consequence of "maintained supersaturation" is that it continually draws attention to the relationship between minerals and the body fluids. This problem has caused much concern among bone physiologists who have invoked osteoclast activity and parathyroid function, renal and intestinal physiology to account for supersaturation. Invertebrate physiologists have never recognized this as a problem and they may very well be right. It seems quite likely that "apparent supersaturation" is an artefact and the reality is that it is the normal situation for unusual minerals in a complex saline. This explains how acidotic animals re-equili-brate at new ionic levels so as to maintain a constant solubility product in a way that is very difficult to imagine by cellular means (but see[635]).

If one now looks at the numerous examples of intracellular calcification that are known, they frequently involve small crystals with concentrically arranged deposits rich in calcium, magnesium, phosphate and carbonate ions. What are the solubility products of these deposits? Is it not likely that they are in equilibrium with the body fluids, i. e., they are impure and have "high" solubility products? When they are released from cells, they will raise the apparent solubility product above that for calcite or aragonite. In fact, if pure minerals are available nearby, the impure crystals may dissolve and recrystallize at those sites in the more stable form. The importance of this argument is, of course, that the energy input and actual production of minerals have occurred within the cell which is therefore the *site of calcification.* When released, however, the crystals dissolve and recrystallize in another region, i. e. *a site of mineralization* (Fig. 11). If this is true, investigations of the site of mineral-ization are unlikely to explain anything about the mechanism of calcification.

4 Mechanisms of Calcification

No final comprehensive hypothesis of the chain of events in biological calcification processes can be constructed at the present time. All current major theories on the sequence of events — leading to translocation of mineral ions into initial mineral phase — involve overcoming the energy of activation required for forming the first primitive stable mineral clusters. Then calcium and phosphate are present at levels

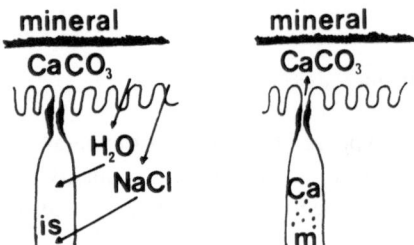

Fig. 11. Two systems in which intercellular spaces may involved in calcification. (After Simkiss[445])

either metastable (or lower) with respect to formation of this initial mineral phase. The most intensively studied theories on mineralization of bone designate the site of this transformation:

(a) Within and upon collagen and other associated matrix components by the process involving heterogeneous nucleation; as well as

(b) within cellular organelles or cell-derived membranes by a process involving calcium uptake or possibly active membrane transport.

The most intensively studied organelles have been the mitochondria. Various enzymatic functions require that intracellular calcium be kept at 10^{-6} to 10^{-7} M in the cytosol and there is firm evidence that mitochondria have an important role in fulfilling this function. Because mitochondria sequester calcium in muscle, kidney and liver, it has been postulated that mitochondria might serve as an apparatus to overcome the energy of activation necessary for forming a new calcium-phosphate mineral phase at calcifying sites[471].

It has been demonstrated that mineral deposition in growth plate cartilage mitochondria[472] in vivo and mitochondrial granules are not only concentrated in cells adjacent to the septa which are about to undergo calcification but contained as well mineral in ashed solutions[473–476]. Low oxygen tension has been regarded as a possible trigger for the onset of calcification[477]. Inhibitors of calcium phosphate crystal growth are found in mitochondria, e. g. Mg^{2+}, nucleotides and citrate. It was postulated that the mineral is probably kept in a non-crystalline form by the presence of one or more inhibitors particularly ATP[471]. The unsolved problems which remain before establishment of the mitochondrial hypothesis as the initial mineral-forming site for extracellular matrices are:

(1) no mitochondria are seen in the extracellular matrix discharging contents; and

(2) no mineral trail extending from mitochondria into cell processes at calcifying sites has yet been found in studies of cartilage, dentine, periosteal bone[474] or mucosal cells of avian shell glands[478].

However, cellular mechanisms for calcium transfer and homeostasis intimately associated with mitochondria and biomineralization are an essential process in Ca^{2+}-regulation[479, 480].

Considerable controversy exists as far as the calcification sites inside or outside the cells are concerned. The nature of this arrangement has been a matter of debate for over two decades since the statement by Neuman and Neuman[481] that " . . . . . . calcification occurs in individuals whose blood levels of calcium and phosphate are well below the critical product". This generated idea of crystal formation is the result of stereotactic properties of some portion of the organic matrix. This view[181, 482] is still maintained and widened by ultrastructural evidence presented in support of the idea of some specific site upon the collagen molecule[483].

The visible evidence in matrix bone cannot be disputed. The "resting osteocyte" is contained within a dense mineral environment, all of which is clearly outside the membrane. It is easy to accept the idea that the matrix is laid down by the osteoblast and subsequently mineralized extracellularly. The dynamic evidence about bone formation, however, is not so clear, since a brief examination of the literature suggests that evidence is so uncertain as to require complete re-examination. The mineral at a developing bone front appears in discrete clusters or pockets with a granular texture verified many times[484] and stated by some authors (e. g.[485]) to be unconnected with the orientation of the fibrous matrix. Many recent reports of calcifying structures of cell origin have served to increase the uncertainty as to where the insoluble mineral originates and to what extent it derives from the cell or from the circulation. Matrix vesicles as described above appear to possess properties of calcification in cartilage and dentine, and have transferred the mineralization event from the "blood levels"[481] to something which is at least confined whithin a membrane at some unknown concentration[486].

However, increasing evidence points to the key role of extracellular matrix vesicles in the initiation of biological calcification[455, 456, 458, 469, 487–490].

Calcification can be visualized in two phases. In the first phase, phosphate accumulates within the matrix vesicles through the enzyme hydrolysis of inorganic pyrophosphate (and perhaps other phosphate esters)[458].

The accumulated phosphate reacts with calcium available to form apatite. Calcium may be concentrated within the vesicles by virtue of a lipid-calcium interaction and possibly by active calcium transport across the vesicle membrane[458]. Recent studies have led to the discovery of PL-Ca-P_i-complexes in calcifying tissues. These Pl-Ca-P_i complexes[110] and certain proteolipids[122, 491] have been shown to rapidly nucleate HA from metastable Ca-PO_4 solutions. There is evidence for the existence of these complexes in matrix vesicles[135].

In the second phase, the crystal within the vesicle grows until it perforates the vesicle membrane and becomes exposed to the cartilage matrix fluid. Then apatite will continue to grow because the cartilage matrix fluid in a normal animal is supersaturated with respect to apatite crystals[492]. This would not be the case in rickets where a low value of (Ca^{2+}) x (PO_4^{3-}) in the cartilage fluid probably explains the failure of the second phase of crystal growth. The rachitic matrix vesicles often contain apatite, although no extravesicular apatite could be detected. This suggests that the vesicles are able to concentrate calcium even in the presence of an unfavourably low ionic product, (Ca^{2+}) x (PO_4^{3-})[458].

In the second stage, as the needle-like calcium phosphate deposits radially grow out of the vesicles, the surrounding collagen fibres become mineralized.

Attempts to combine stereotaxis with the nature of calcifying vesicles leads to a no more satisfactory explanation of what happens; for either the calcification site is inside the vesicle or outside it. If it is inside, the role of collagen is irrelevant, and if it is outside on the collagen, the role of the vesicles is irrelevant. There is no evidence that the mineral is, *ab initio,* inside the vesicle. If so, then the mineralization event would be within the membrane, and by inference, within the cell that produced it. The situation is still further confounded by opposing experimental explanations. The first is that there may be two mechanisms with two distinct pathways of calcification, the one within a membrane, the other freely in equilibrium with the serum fluids. For this there is no strong evidence either way. The second is that mineralization does occur within a vacuolar membrane both inside and outside the cell, but the early stages of the process are too soluble or too easily disrupted to be retained for visualization by conventional techniques. It has been assumed that the second of these possibilities may be correct and that mineralization in bone is analogous to that in *Spirostomum* and the site of formation in both cases is the Golgi apparatus[486].

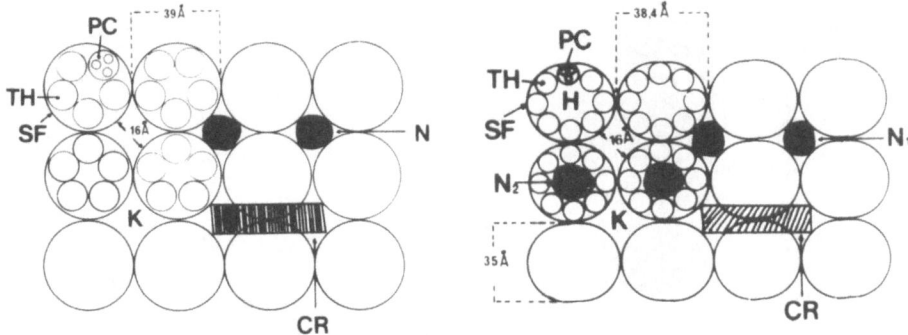

Fig. 12. Diagrammatical cross section through models of collagen mineralization. **a)** Tetragonal arrangement of subfibrils in the collagen fib. according to Miller and Parry[643]. Since subfibrils (SF) are helically wound no hole regions but channels (K) exist between the subfibrils. There may be the locales for initiation of mineral formation (N). PC = protein chain; Th = Triple helix; CR = apatite crystal; K = channels between subfibrils; **b)** Orthorhombic model. (H = hole region in subfibrils; N_1 = mineral nuclei between subfibrils; N_2 = mineral nuclei within a subfibril.) (After Höhling et al.[641])

However, it has been found in two hard tissue systems that protein-polysaccharide-rich regions are mineralized to a higher degree than collagen-rich regions[493]. The first mineral deposits are no longer predominantly associated with vesicles, but are found without any significant time difference in the ground substance and collagen[494]. For these reasons and because of the great importance of collagen to the tissue structure, one should not belittle the role of collagen mineralization. Höhling[495−499] came to the conclusion that dot-like nuclei were initially formed on and within the collagen fibres orientated along the fibre axis (Fig. 12a, Fig. 12b). At first they form chains and then needles by fusion along this axis and later on often plate-like crystals by recrystallization. It was assumed that the distances between the point nuclei mirrored the distances between nucleation centres on the collagen molecules. In the direction of the long axis, the distances between the dot-like nuclei were in good agreement with the gross-banding pattern of the collagen. Also the number of nuclei per collagen macroperiod approximates the number of cross-striations and therefore particularly the amino acid sequences of the primary structure. However, the relationship between the lateral separation of the calcium phosphate needles and the macromolecular structure of collagen is not so clear. Recent X-ray analyses of tendon collagen have shown, that the smallest morphological subunits are not triple helices but subfibrils, built from five triple helices[500−502]. Model building based on the indices corresponding to the diffraction pattern led to the conclusion[502] that the tetragonal arrangement of the subfibrils most clearly fitted the corresponding low angle reflection indices.

The subfibrils with diamters of 39 Å contain five helices each of which in turn is built from three polypeptide chains. In this arrangement, microchannels or holes having minimum diameters of approximately 16 Å are enclosed between the subfibrils.

The five triple helices of the subfibrils are coiled around each other to form another large helical structure[503], and thus pronounced microchannels are not found within the subfibrils.

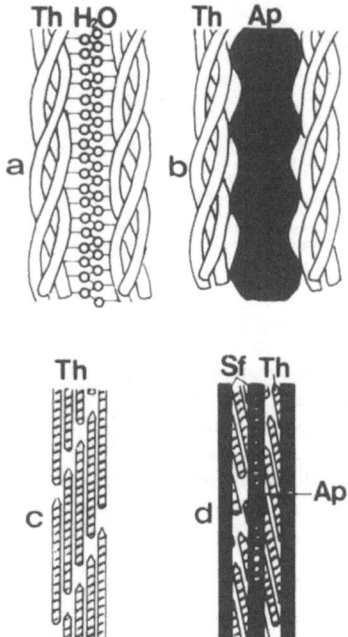

Fig. 13. Longitudinal section through models of collagen mineralization. **a)** Triple helix (Th) with water molecules in native state; **b)** Triple helix with apatite nuclei (Ap) (not very likely); **c)** Two-dimensional presentation of 0.4 D-final overlappings of 5 triple helices; **d)** Aggregation of 5 triple helices to form subfibrils (Sf) and formation of apatite nuclei between subfibrils. (After Höhling et al.[641])

Regarding the location within the collagen fibre where the mineral nuclei are expected to be formed, it seems that the most likely sites are the microchannels between the subfibrils. If one assumes that the subfibrils are directly touching and thus lying in dense arrays, the average distance between mineral nuclei in the channels would be about 39 Å. It must be assumed in this model that the onset of mineralization results in the widening of the separation between subfibrils, and eventually in a loss of contact between neighbouring subfibrils. The growth of the nuclei and their fusion to form crystals accelerates the process of fibre enlargement (Fig. 13).

From the assumption that the collagen fibres are expanded, it also follows that the diameter values of calcium phosphate nuclei should exceed the value of 16 Å, which is the minimum diameter of the microchannels (i. e. the diameter when the subfibrils are in direct contact).

On the basis of this model it is also possible to propose that the first mineral nuclei do not touch because they are separated by subfibrils, and that they gradually grow and unite to form crystals by separating the non-covalent bonds between neighbouring subfibrils[483].

4.1 Mechanisms of Action of Proteoglycan Inhibition

It has been postulated that proteoglycans play an important role on mineral growth[253, 255, 504, 505]. Also, a specifically mineral-generative role for these compounds has been suggested[506–508]. Evidence has been found for calcium storage in puppy and calf growth cartilage, respectively[509, 510], prior to calcification. Chondroitin sulfate alone was not the agent binding calcium[511, 512]. No calcium binding or phosphate binding to proteoglycan aggregates or subunits in the presence of a synthetic lymph with sodium-to-calcium ratio (50 : 1) could be observed[528]. It has been suggested that inhibition by proteoglycans is directed against already formed mineral clusters, rather than preventing cluster formation[513]. This view is based on a current concept of proteoglycan structure involving aggregate and subunit forms of these molecules discussed elsewhere[514–519].

The fact that simple physical phenomena were involved in the inhibition of mineral growth in a solution containing calcium and phosphate, supersaturated with respect to the nascent phase, was interpreted to mean that shielding forces at the surface of the embryonic mineral clusters were provided by proteoglycan aggregates but not by the subunits[520]. It has been postulated that only the fastsedimenting fraction of aggregates has sufficient shielding properties to sequestered mineral clusters[521]. Disaggregation of the proteoglycan aggregates or simply dilution of the adjacent molecules might reduce or remove the inhibitory effects of the large aggregates[521]. Once needles of apatitic crystal are found, their aggregation seems to be potentially inhibited by proteoglycan aggregates[522].

Experimental treatment with proteolytic enzymes of growth cartilage matrices displayed narrow growth plate with extension of calcification into the proliferating cell zones in comparison to controls[523]. There is also direct evidence of proteoglycan catabolism in calcifying growth cartilage[261, 513].

Several investigations established a role for lysosomal proteases, particularly cathepsin D, in degradation of cartilage [524–528]. However, in articular cartilage a metal-dependent, protease probably not lysosomal-neutral, has also been identified[529, 530]. Addition of this cartilage enzyme caused rapid degradation of proteoglycan subunit preparations[513].

Calcification occurs in normal cartilage at the metaphyseal junction. The greatly expanded histological zone in healing rickets proximal to the metaphyseal junction yields data which have been interpreted to mean the presence of a preliminary reversible step of partial proteoglycan disaggregation, following which inhibition of mineral growth disappears. The agency of this first step might comprise an enzymic action not yet identified which is directed against certain parts of the proteoglycan aggregate, e. g. hyaluronic acid[513]. Free radicals generated from lipid peroxides[531] might act to dismember hyaluronates.

The action of lysozyme for aggregation and disaggregation of proteoglycans, action alone or in concert with an unknown agency studied in vitro systems, produced highly reproducible data[521]. If lysozyme is functional (and not a concurrent fortuitous event), lysozyme could regulate proteoglycan aggregates through altered concentrations, opening then theoretically to a subsequent degradative enzymatic attack, and thereby controlling certain physical properties of the matrix. It should be emphasized that whereas evidence of this function of lysozyme is lacking, no other regulation of proteoglycan aggregation size among many agents tested has been found which fulfills the requirements of reversibly reducing aggregation[513].

4.2 Microinhibition of Mineral Formation

Another possible role for inhibition of mineral growth by pyrophosphate and possibly nucleotides has been discussed in detail elsewhere[114].

The inhibition could involve elaboration and adherence to mineral of pyrophosphate and enzymatic removal of the factor (e. g. soluble factor as pyrophosphate ion) or could include incorporation of the inhibitor into the mineral phase as may occur in mitochondrial mineral formation[114]. ATP[114], pyrophosphate[532], and diphosphoglycerate[532] may all have such a role in calcifying sites. These substances could function to control mineral at first formed in membraned particle sites or nucleational sites, be hydrolyzed, and the proteoglycan inhibitor role could comprise a second step in the process. In this connection, it was shown that proliferation and hypertrophic cell zone cartilage, but not resting cells, elaborate pyrophosphate and alkaline phosphatase into the incubation medium[513]. Inhibition of alkaline phosphatase with dithiothreitol and cysteine inhibited the output of pyrophosphate which is non-dialyzable, but could be separated from large molecules[513]. It has also been reported that pyrophosphate at very low concentrations may stimulate mineral production by matrix vesicles[458].

4.3 Effects of Diphosphonates on Calcification

Diphosphonates are pyrophosphate analogues, and they are characterized by P-C-P-linkage instead of P-O-P bonds. They have been shown to have profound effects on calcification of tissues[533–535]. Of the many diphosphonates, ethane-1-hydroxy-1,1-diphosphonate (EHDP) and dichloromethylene diphosphonate (CL_2MDP) appear to be the most biologically active compounds[536]. EHDP and Cl_2MDP bind to HA with EHDP having a greater binding capacity[537]. Both diphosphonates retard the dissolution of HA in vitro[536, 538, 539]. Cl_2MDP was found to be more effective than EHDP in inhibiting bone resorption in vitro[536, 539, 340] and in vivo[540–542]. Higher doses of EHDP inhibited the mineralization of bone[543], whereas this effect has not been observed after Cl_2MDP treatment. However, Cl_2MDP is just as effective as EHDP in preventing the calcification of soft tissues[544]. Collagen catabolism is depressed to a greater extent than collagen synthesis[547]. These findings are in accord with other investigations[535, 542, 545–547]. Osteoclast population parameters are increased with both Cl_2MDP and EHDP with Cl_2MDP having a greater effect than similar doses of EHDP[548].

It has been shown that diphosphonates can affect cellular activity and that these changes, particularly those in local production, can account for the actions of Cl_2MDP and EHDP on bone resorption[550].

The effect of diphosphonates on biomineralization processes equals that of pyrophosphate and long-chain polyphosphates. It has been assumed that diphosphonates inhibit formation and growth of crystals of calcium phosphate by binding to the nuclei and already formed crystals[535]. The possibility of binding of phosphate and condensated phosphates to collagen or of blocking nucleation sites of collagen or of structural modification of nucleation sites as well as allosteric effects has been discussed[535].

However, methane-diphosphonate could not prevent the growth of apatite crystals in vitro on prepared sinews of rats tail out of a metastable solution with calcium and phosphate ions. On the contrary, the precipitated crystalline particles were bigger and better crystallized than those from control solutions. This is in surprising contrast to most of the information from the literature. No other calcium phosphate minerals besides apatite have been found by X-ray diffraction, whereas under comparable conditions brushite and octacalcium phosphate grow on collagenous sinews[549].

Treatment of animals with large doses of EHDP is known to reduce the mineralization of bone[542], calcium absorption in the intestine[541, 551, 552] and 1.25-dihydroxycholecal-ciferol production by the kidney[553–555].

The administration of a small dosis of 1.25-dihydroxycholecalciferol normalizes calcium absorption, but not bone mineralization[556, 557]. EHDP might inhibit the renal 1-hydroxylase directly[553, 558]. The EHDP-induced inhibition of 1.25-$(OH)_2D_3$ production has been shown to be reduced by a low Ca diet or by vitamin D deficiency[554, 555]. The influence of EHDP on the renal 1-hydroxylation is indirect and dependent on dietary vitamin D, calcium, and phosphorus[559].

4.4 Calcification Processes in Single Cells

In single-celled organisms, there is no escape from the conclusion that the cell itself must somehow be involved in the transport and assembly of the inorganic ions into structures which mature outside or inside the cytoplasm. The organism is completely in control of the entire process of mineralization, directly or indirectly. There are no complex cell interactions and divisons of labour to abscure the contributions by organelles. There is no intercellular matrix with properties which may be different from the cells within it. There is no enclosed vascular system to confuse information on the concentration and location of ions at any given time.

A definition of "calcification" in single cells is difficult. The accumulations of inorganic salt may be sufficiently large and crystalline to present an unmistakeable mass, but there are many less well-defined arrangements where the calcium salts are diffuse and transitory, yet stable and unchanging over periods long enough to be regarded as parts of the cell fabric[486].

For the purpose of relating calcium deposits in unicellular organisms to mineralized metozoan tissue, a broad division might be made between those calcium-rich regions which have no fixed morphological identity and those structures which contain inorganic matter as part of a more permanent, repeated pattern of some kind. The deciding factor is whether the calcium moiety is mobile and part of the transport system or wether it is not readily changed in nature or position for periods of time and becomes a specific part of the cell or its components. A definition of this kind would embrace microtubules and even the nucleus. The mineral is less important than its morphological site and constancy[486].

Perhaps the most consistent feature of the calcified structures in unicellular organisms is their characteristic shape. Virtually all the objects that are mineralized have definite boundaries within which the inorganic component is maintained to a constant degree of composition and crystallinity. The Foraminifera are a good example of such controlled deposits within boundaries[486].

4.4.1 Coccoliths

Coccoliths, and the related scales and sheaths, are characteristic features of a widespread group of organisms. The morphology of each structure is typical of each species and represents a range of extracellular objects from simple mineral-free plates to intrically sculptured, calcified and crystalline units fitted together to make a wreath or incrustation around the cell (for reviews see[560, 561]).

In some species, the calcite crystals arise outside the plasmalemma on templates produced in the cytoplasm[562]. In species such as *Coccolithus huxleyi*, however, the mineralized placoliths are made within the cell and extruded[563]. More recent observations now leave less room for doubt that the coccoliths in this organism are made entirely within the saccules of the Golgi apparatus by a series of events. The organic framework is not constructed within the Golgi system, as might be expexted (e. g. *Paraphysomonas vestita* scales[564] and *Pleurochrysis* wall formation[565]). A similar cycle of events is likely for *Coccolithus huxleyi*[563], see also[486].

In the coccolithophorids, then, there is accumulating evidence for various degrees of cell involvement in biomineralization: manufacture of uncalcified scales which remain uncalcified, scale frameworks which calcify outside the cell membrane, and coccoliths which are made and calcified within the cell. In each case, the Golgi apparatus plays the principal role in the fabrication, and the product is morphologically distinct and characteristic for the organism[456]. Calcium phosphate as "bone salt" is present within the ciliate *Spirostomum ambiguum*[566]. The mineral in *Spirostomum* is arranged in the form of round, or nearly round, particles, each composed of clusters of filaments, formed in vacuoles within the cell. In bone, the mineral is mostly in the form of small crystals associated with collagen outside the cell[486]. The mineral may be maintained in domains of some kind in relation to an organic envelope. The particles are extruded in packets under duress; otherwise they are retained within the cytoplasm. The assemblage of filaments is made within vacuoles in association with the nucleus which may be highly modified[486]. To what extent *Spirostomum* goes through the extrusion phase of the osteocyte at some stage of its life cycle and to what extent the osteocyte requires movable cytoskeletal particles analogous to those in the protozoan, remains to be seen. At this stage it is only clear that the calcium phosphate in both the unicellular organism and the independent vertebrate cell is packaged by complex process and has the function required of it[486].

The present studies on calcium phosphate in bacteria are mostly of dental interest. Through the mineralization of the dental plaque dental calculus forms the soft, adherent and predominant coating which forms on the surface of teeth[567]. X-ray diffraction studies have shown the dental calculus to be composed of four principal minerals: hydroxyapatite, octacalcium phosphate, brushite and whitelockite[568]. The mechanism by which mineralization of oral calculi is initiated is not fully understood[567]. Two types of mineralization centres can be distinguished which apparently constitute different mechanisms of formation. These have been called centres of mineralization "types A und B". Mineralization centres of type A are initiated and formed only in the presence of, and in association with, microorganisms[568]. Type B has not been studied thoroughly sofar. The effects of phosphonates on calculus formation has been discussed elsewhere[569, 570].

5 References

1. Lowenstam, H. A.: Biologic problems relating to the composition and diagenesis of sediments. In: The earth sciences: problems and progress in current research, pp. 137. Donelly, T. W. (ed.). Chicago: Chicago Press 1963
2. Lowenstam, H. A.: Biogeochemistry of hard tissues, their depth and possible pressure relationships. In: Barobiology and the experimental biology of the deep sea, pp. 19. Brauer, R. W. (ed.). Chapel Hill: Univ. North Carolina 1973
3. Degens, E. T.: Molecular Mechanisms on Carbonate, Phosphate, and Silica Deposition in the Living Cell. Topics Curr. Chem., pp. 1. Boschke, F. L. (ed.). Springer: Berlin Heidelberg New York 1976
4. Brooks, R., Clark, L. M., Thorston, E. F.: Calcium carbonate and its hydrates. Phil.Trans. Ray.Soc. (London) A *243,* 145 (1951)
5. Bütschli, O.: Untersuchungen über organische Kalkgebilde. Abh. kgl. Ges. Wiss. Göttingen, math.-phys. Kl. N. F. J. 6, Nr. 3 (1908)
6. Johnston, J., Mervin, H. E., Williamson, E. D.: The several forms of calcium carbonate. Amer. J. Sci. *41,* 473 (1916)
7. Linck, G.: Über die Bildung der Carbonate des Calciums, Magnesiums und Eisens. In: Doelters Handb. Mineralchem. *1,* 113 (1911)
8. Lippmann, F.: Versuche zur Aufklärung der Bildungsbedingungen von Calcit und Aragonit. Fortschr. Min. *38,* 156 (1960)
9. Mayer, F. K.: Röntgenographische Untersuchungen über die Modifikationen des Kalziumkarbonates in Gastropodenschalen. Chemie d. Erde *6,* 239 (1931)
10. Mayer, F. K., Weineck, E.: Die Verbreitung des Kalziumkarbonates im Tierreich unter besonderer Berücksichtigung der Wirbellosen. Jenaische Z. Naturwiss. *66,* 199 (1932)
11. Meyer, H. J.: Bildung und Morphologie des Vaterits. Z. Kristallogr. *121,* 220 (1965)
12. Meyer, H. J.: Struktur und Fehlordnung des Vaterits. Z. Kristallogr. *128,* 138 (1969)
13. Moda, T.: Crystals of Calcium carbonate. J. Soc. Chem. Ind. Japan *37,* (Suppl.) B. 319 (1934)
14. Spangenberg, K.: Die verschiedenen Modifikationen des Calciumcarbonates. Z. Kristallogr. *56* 432 (1922)
15. Bischoff, J. L. and Fyfe, W. S.: Catalysis inhibition and the calcite-aragonite problem. I. The aragonite-calcite transformation. Amer. J. Sci. *266,* 65 (1968)
16. Chave, K. E.: Aspects of the biogeochemistry of magnesium. 1. Calcareous marine organism. J. Geol. *62,* 266 (1954)
17. Chave, K. E., Deffeyes, K. S., Weyl, P. K., Garrels, R. M., Thompson, M. E.: Observations on the solubility of skeletal carbonate in aqueous solutions. Science *137,* 33 (1962)
18. Kitano, Y.: Chemistry of fossils - Factors controlling the crystal form and the minor element contents of carbonates formed in biological systems. Chemy Chem. Ind. Tokyo *21,* 1017 (1968)
19. Kitano, Y., Kanamori, N., Tokuyama, A.: Effects of organic matter on solubilities and crystal form of carbonates. Amer. Zool. *9,* 681 (1969)
20. Kitano, Y., Kanamori, N., Tokuyama, A.: Influence of organic matter on inorganic precipitation. In: Organic matter in natural waters, pp. 413. Hood, D. W. (ed.). Inst. Marine Sci., Univ. Alaska, 1970

21. Kitano, Y., Kanamuri, N., Tokuyama, A., Omorı, T.: Factors controlling the trace-element contents of marine carbonate skeletons. In: Proceedıngs of Symposium on Hydrogeochemıstry and Biogeochemistry (Vol. 1-Hydrogeochemistry), pp. 484. Washington, D. C.: The Clarke Co. 1973

22. Kitano, Y., Yoshioka, S., Kanamori, N., Tsuzuki, T.: The Transformatıon of Aragonite to Calcite in Aqueous Solutions. Fossils, *23–24*, 15 (1972)

23. Milliman, J. D.: Marine carbonates. New York, Heidelberg, Berlin: Springer 1974

24. Watabe, N., Wilbur, K. M.: Influence of the organic matrix on crystal type in molluscs. Nature *188*, 334 (1960)

25. Wilbur, K. M.: Shell Structure and Mineralization in Molluscs. In: Calcification in biological systems, pp. 15. Sognnaes, R. F. (ed.). Amer. Assoc. Adv. Sci., Washington 1960

26. Kitano, Y., Kanamori, N., Yoshioka, S.: Influence of chemıcal specıes on the crystal type of calcıum carbonate. In: The mechanisms of mineralization in the invertebrates and plants, pp. 191. Watabe, N. and Wilbur, K. M. (ed.). The Belle W. Baruch Library in Marine Science Vol. 5, University of South Carolina Press, 1976

27. Comar, C., Bronner, F. (eds.): Mineral metabolism. New York: Academic Press 1969

28. Erben, H. K.: Ultrastrukturen und Mineralisation rezenter und fossiler Eischalen bei Vögeln und Reptilien. Biomineralisation *1*, 1 (1970)

29. Erben, H. K.: Über die Bıldung und das Wachstum von Perlmutt. Biomineralisation *4*, 16 (1972)

30. Mc Connell, D.: Apatite. Its crystal chemistry, mineralogy and biological occurrences. Wien-New York: Springer 1973

31. Mc Lean, F. C., Urist, M. R.: Fundamentals of the Physiology of skeletal tissue. Chicago: University Press 1968

32. Mc Lean, F. C., Urist, M. R.: Calc. Tiss. Res. *1*, 1 (1967)

33. Mutvei, H.: On the micro- and ultrastructure of the conchiolin in the nacreous layer of some recent and fossil molluscs. Acta Univer. Stockholmiensis, Stockholm Contr. Geol. *20*, 1 (1969)

34. Mutveı, H.: Ultrastructural relationships between the prismatic and nacreous layers in Nautilus (Cephalopoda). Biomineralisation *4*, 81 (1972)

35. Mutvei, H.: Ultrastructure of the mineral and organic components of molluscan nacreous layers, Biomineralisation *2*, 48 (1970)

36. Wise, S. W.: Microarchitecture and mode of formation of nacre (mother of pearl) in pelecepods, gastropods and cephalopods. Eclogae geol. Helv. *63*, 775 (1976)

37. Zipkin, I. (ed.): Biological Mineralization. New York-London-Sidney-Toronto: John Wiley 1973

38. Matheja, J., Degens, E. T.: Structural molecular biology of phosphates. Stuttgart: Gustav Fischer Verlag 1971

39. Neuhaus, A.: Orientierte Kristallabscheidung (Epitaxie). Angew. Chem. *64*, 158 (1952)

40. Armstrong, W. D., Singer, L.: Composition and constitution of the mineral phase of bone. Clin. Orthop. *38*, 179 (1965)

41. Garrels, R. M., Crist, C. L.: Solutions, minerals, and equilibria, New York: Harper and Row 1965

42. Mitterer, R. M.: Biogeochemistry of aragonite mud and oolites. Geochim. Cosmochim. Acta *36*, 1407 (1972)

43. Tooms, J. S., Summerhayes, C. P., Cronan, D. S.: Geochemistry of marine phosphate and manganese deposits. Oceanogr. Mar. Biol. Ann. Rev. *7*, 49 (1969)

44. Termine, J. D., Eanes, E. D.: Comparative chemistry of amorphous and apatitic calcium phosphate preparations. Calc. Tiss. Res. *10*, 17 (1972)

45. Selvig, K. A.: Periodic lattice images of hydroxyapatite crystals in human bone and dental hard tissues. Calc. Tiss. Res. *6*, 227 (1970)

46. Selvig, K. A.: The crystal structure of hydroxyapatite in dental enamel as seen with the electron microscope. J. Ultrastructural Res. *41*, 369 (1972)

47. Selvig, K. A.: Electron microscopy of dental enamel: Analysis of crystal lattice images. Z. Zellforsch. *137*, 271 (1973)

48. Münzenberg, K. J.: Untersuchungen zur Kristallographie der Knochenminerale. Biomineralisation *1*, 67 (1970)

49. Posner, A. S., Perloff, A., Diorio, A. F.: Refinement of the hydroxyapatite structure. Acta Cryst. *11*, 308 (1958)

50. Kay, M. I., Young, R. A., Posner, A. S.: Crystal structure of hydroxyapatite. Nature *204*, 1050 (1964)

51. Young, R. A., Elliot, J. C.: Atomic-scale bases for several properties of apatites. Arch. Oral Biol. *11*, 699 (1966)

52. Wondratschek, H.: Z. Kristallchemie des Blei-Apatits (Pyromorphit). Neues Jb. Miner. Abh. *99*, 113 (1963)

53. Bauer, H.: Über eine Apatit-artige Verbindung der Formel $Ba_{10} (PO_4)_5 (BO_4)$, Angew. Chem. *71*, 374 (1959)

54. Stevenson, J. S., Stevenson, L. S.: Fluorine content of microsaur teeth from the carboniferous rocks of Joggins, Nova Scotia. Science *154*, 1548 (1966)

55. Posner, A. S., Perloff, A.: Apatites deficient in divalent cations. J. Res. Nat. Bur. Stand *58*, 279 (1957)

56. Posner, A. S., Eanes, E. D., Harper, R. A., Zipkin, I.: X-ray diffraction analysis of the effect of fluoride on human bone apatite. Arch. Oral Biol. *8*, 549 (1963)

57. Engström, A., Zetterström, R.: Studies on the ultrastructure of bone. Exp. Cell. Res. *2*, 268 (1951)

58. Finean, J. B., Engström, A.: The low-angle scatter of x-rays from bone tissue. Biochim. Biophys. Acta *11*, 178 (1953)

59. Molnar, E.: Additional observations on bone crystal dimensions. Clin. Orthop. *17*, 38 (1960)

60. Robinson, R. A., Watson, M. L.: Crystal and collagen morphology as a function of age. Ann. N. Y. Acad. Sci. *60*, 596 (1955)

61. Correy, J. D.: Three analogies to explain the mechanical properties of bone. Biorheology *2*, 1 (1969)

62. Levine, P. T., Glimcher, M. J., Seyer, J. M., Huddelston, J. I., Hein, J. W.: Non-collagenous nature of the proteins of shark enamel. Science *154*, 1192 (1966)

63. Frazier, P. D.: X-ray diffraction analysis of human enamel containing different amounts of fluoride. Arch. Oral. Biol. *12*, 35 (1967)

64. Ascenzi, A., Bonucci, E.: The osteon calcification as revealed by the electron microscope. In: Calcified Tissues 1965, Proc. Third European Symp. on Calcified Tissued, pp. 142–146. Fleisch, H., Blackwood, H. J. J., and Owen, M. (ed.). New York, Heidelberg, Berlin: Springer 1966

65. Menczel, J., Posner, A. S., Harper, R. A.: Age changes in the crystallinity of rat bone apatite. Israel J. Med. Sci. *1*, 251 (1965)

66. Mielsen, A. E.: Kinetics of precipitation Chap. 9. New York: Macmillan 1964

67. Eastoe, J. E.: The chemical Composition of Bone. In: Biochemist's Handbook, pp. 715. Long, C. (ed.). Princeton, N. J.: Van Nostrand Ca. 1961

68. Woodard, H. Q.: The elementary composition of human cortical bone. Health Phys. *8*, 513 (1962)

69. Neuman, W. F.: Collected Studies on Hydroxyapatite. The University of Rochester Atomic Energy Project. Rep. No. UR-238. Rochester: Univ. Rochester 1953

70. Winand, L.: Physico-chemical study of some apatitic calcium phosphates. In: Tooth enamel, pp. 15. Stack, M. V., Fearnhead, R. W. (eds.). Bristol: John Wright and Sons 1965

71. Brown, W. E.: Crystal Growth of Bone Mineral. Clin. Orthop. *44*, 205 (1966)

72. Newesely, H.: Die Realstruktur von Oktacalciumphosphat. M. Chemie, *95*, 94 (1964)

73. Katz, S., Beck, C. W., Muhler, J. C.: Crystallographic evaluation of enamel from caries and noncarious teeth. J. Dental Res. *48*, 1280 (1969)

74. Füredi-Milhofer, H., Purgaric, B., Brecevic, Lj., Pavkovic, N.: A study of the precipitates in the physiological pH region. Calc. Tiss. Res. *8*, 142 (1971)

75. Brecevic, Lj., Füredi-Milhofer, H.: Precipitation of calcium phosphates from electrolyte solutions. II. The formation and transformation of the precipitates. Calc. Tiss. Res. *10*, 82 (1972)

76. Brown, W. E., Smith, J. P., Lehr, J. R., Frazier, A. W.: Crystallographic and chemical relations between octa-calcium phosphate and hydroxyapatite. Nature *196*, 1050 (1962)
77. Eanes, E. D., Posner, A. S.: Kinetics and mechanisms of conversion of noncrystalline calcium phosphate to crystalline hydroyapatite. Trans. N. Y. Acad. Sci. Ser. II. *28*, 233 (1965)
78. Eanes, E. D., Termine, J. D., Posner, A. S.: Amorphous calcium phosphate in skeletal tissues. Clin. Orthop. *53*, 223 (1967)
79. Termine, J. D., Peckauskas, R. A., Posner, A. S.: Calcium phosphate in vitro. II. Effects of environment of amorphous-crystalline transformation. Arch. Biochem. Biophys. *140*, 318 (1970)
80. Newesely, H.: Kristallchem. Argumente zur Kariesprophylaxe durch Fluoridierungsmaßnahmen. Dtsch. zahnärztliche Ztschr. *24*, 1483 (1967)
81. Zimmerman, S.: Physiochemical properties of Enamel and Dentine. In: Dental biochemistry, pp. 112. Lazarri, E. P. (ed.). Philadelphia: Lea & Febiger 1976
82. Dosch, W., Koestel, C.: Rasterelektronenmikroskopie von Harnsteinen. In: Fortschr. Urologie u. Nephrologie. Bd. 7, Pathogenese und Klinik der Harnsteine, pp. 41. Vahlensieck, W., Gassner, G. (eds.). Darmstadt: Steinkopf Verlag 1975
83. Molnar, Z.: Development of the parietal bone of young mice. I. Crystals of bone mineral in frozen-dried preparations. J. Ultrastructal Res. *3*, 39 (1959)
84. Eanes, E. D., Gillessen, I. H., Posner, A. S.: Mechanism of conversion of non-crystalline calcium phosphate to crystalline hydroxyapatite. In: Crystal Growth, pp. 373. Peiser, H. S. (ed.). Oxford: Pergamon Press Inc. 1967
85. Eanes, E. D., Harper, R. A., Gillessen, I. H., Posner, A. S.: An amorphous component in bone mineral. In: The Fourth European Symp. Calcified Tissues, pp. 24. Gaillard, P. J., Van den Hoof, A., Steendijk, R. (eds.). Amsterdam: Exerpta Medica 1966
86. Eanes, E. D., Posner, A. S.: Structure and chemistry of bone mineral. In: Biological calcification: cellular and molecular aspects, pp. 1. Schraer, H. (ed.). Amsterdam: North Holland Publishing Co. 1970
87. Eanes, E. D., Termine, J. D., Nylen, M. U.: An electron microscope study of the formation of amorphous calcium phosphate and its transformation to crystalline apatite. Calc. Tiss. Res. *12*, 144 (1973)
88. Schraer, H., Gay, C. V.: Matrix vesicles in newly synthesizing bone observed after ultra-cryotomy and ultramicroincineration. Calc. Tiss. Res. *23*, 185 (1977)
89. Boskey, A. L., Posner, A. S.: Conversion of amorphous calcium phosphate to microcrystalline hydroxyapatite: pH-dependent, solution-mediated, solid-solid conversion. J. Phys. Chem. *77*, 2313 (1973)
90. Watson, M. L., Robinson, R. A.: Collagen-crystal relationships in bone. II. Electron microscope study of basic calcium phosphate crystals. Amer. J. Anat. *93*, 25 (1953)
91. Termine, J. D., Posner, A. S.: Infrared analysis of rat bone: Age dependency of amorphous and crystalline mineral fractions. Science *153*, 1523 (1966)
92. Termine, J. D., Posner, A. S.: Calcium phosphate formation in vitro. I. Factors affecting initial phase separation. Arch. Biochem. Biophys. *140*, 307 (1970)
93. Fleisch, H., Russell, R. G. G., B. Saz, S., Termine, J. D., Posner, A. S.: Influence of pyrophosphate on the transformation of amorphous to crystalline calcium phosphate. Calc. Tiss. Res. *2*, 49 (1968)
94. Howell, D. S., Pita, J. C., Marquez, J. F., Gatter, R. A.: Demonstration of macromolecular inhibitor (s) of calcification and nucleational factor (s) in fluid from calcifying sites in cartilage. J. Clin. Invest. *48*, 630 (1969)
95. Pita, J. C., Cuervo, L. A., Madruga, J. E., Muller, F. J., Howell, D. S.: Evidence for a role of proteinpolysaccharides in regulation of mineral phase separation in calcifying cartilage. J. Clin. Invest. *49*, 2188 (1970)
96. Cotmore, J. M., Nichols, G., Wuthier, R. E.: Phospholipid-calcium phosphate complex. Enhanced calcium migration in the presence of phosphate. Science *172*, 1339 (1971)
97. Wuthier, R. E., Bisaz, S., Russell, R. G. G., Fleisch, H.: Relationship between pyrophosphate, amorphous calcium phosphate and other factors in the sequence of calcification in vivo. Calc. Tiss. Res. *10*, 198 (1972)

98. Irving, J. T., Wuthier, R. E.: Histochemistry and biochemistry of calcification with special reference to the role of lipids. Clin. Orthop. *56*, 237 (1968)

99. Wuthier, R. E.: Lipids of mineralizing epiphyseal tissues in the bovine fetus. J. Lipid Res. *9*, 68 (1968)

100. Eisenberg, E., Wuthier, R. E., Frank, R. B., Irving, J. T.: Time study of in vivo incorporation of ^{32}P orthophosphate into phospholipids of chicken epiphyseal tissues. Calc. Tiss. Res. *6*, 32 (1970)

101. Wuthier, R. E.: Zonal analysis of phospholipids in the epiphyseal cartilage and bone of normal and rachitis chickens and pigs. Calc. Tiss. Res. *8*, 36 (1971)

102. Peress, N., Anderson, H. C., Sajdera, S. W.: The lipids of matrix vesicles from bovine fetal epiphyseal cartilage. Calc. Tiss. Res. *14*, 275 (1974)

103. Wuthier, R. E.: Enzymatic, lipid, and electrolyte composition of epiphyseal cartilage subcellular fractions. Int. Ass. Dent. Res., 51 st. Gen. Session, Abstr. No. 468, p. 36 (1973)

104. Nash, H. A., Tobias, J. M.: Phospholipid membrane model: Importance of phosphatidyl serine and its cation exchanger nature. Proc. Nat. Acad. Sci. (Wash.) *51*, 476 (1964)

105. Abramson, M. B., Katzman, R., Gregor, H. P.: Aqueous dispersions of serine. Ionic properties. J. Biol. Chem, *239*, 70–76 (1964)

106. Hendrickson, H. S., Fullington, J. G.: Stabilities of metal complexes of phospholipids: Ca (II), Mg(II) and Ni (II) complexes of phosphatidyl serine and triphosphoinositide. Biochemistry *4*, 1599 (1965)

107. Joos, R. W., Carr, C. W.: The binding of calcium to mixtures of phospholipids. Proc. Soc. exp. Biol. (N. Y.) *124*, 1268 (1967)

108. Wuthier, R. E., Eanes, E. D.: Effect of phospholipids on the transformation of amorphous calcium phosphate to hydroxyapatite in vitro. Calc. Tiss. Res. *19*, 197 (1975)

109. Boskey, A. L., Posner, A. S.: Extraction of a Calcium-Phospholipid-Phosphate Complex from Bone. Calc. Tiss. Res. *19*, 273 (1976)

110. Boskey, A., Posner, A. S.: The role of synthetic and bone extracted Ca-Phospholipid-PO_4 complexes in hydroxyapatite formation. Calc. Tiss. Res. *23*, 251 (1977)

111. Miller, A. G., Burnell, J. M.: The Effect of Crystal Size Distributions on the Crystallinity Analysis of Bone Mineral. Calc. Tiss. Res. *24*, 105 (1977)

112. Eanes, E. D., Gillessen, I. H., Posner, A. S.: Intermediate stages in the precipitation of hydroxyapatite. Nature *208*, 365 (1965)

113. Lehninger, A. L.: Mitochondria and calcium ion transport. Biochem. J. *119*, 129 (1970)

114. Betts, F., Blumenthal, N. C., Posner, A. S., Becker, G. L., and Lehninger, A. L.: Atomic structure of intracellular amorphous calcium phosphate deposits. Proc. Nat. Acad. Sci. (Wash.) *72*, 2088 (1975)

115. Becker, G. L., Chen, C., Greenwalt, J. W., Lehninger, A. L.: Calcium phosphate granules in the hepatopancreas of the blue crab *Callinetes sapidus*. J. Cell. Biol. *61*, 316 (1974)

116. Lehninger, A. L.: The mitochondrion, p. 163. New York: W. A. Benjamin, 1964

117. Blumenthal, N. C., Betts, F., Posner, A. S.: Nucleotide stabilization of amorphous calcium phosphate. Mat. Res. Bull. *10*, 1055 (1975)

118. Boskey, A. L., Posner, A. S.: Magnesium stabilization of amorphous calcium phosphate: A kinetic study. Mat. Res. Bull. *9*, 907 (1974)

119. Blumenthal, N. C., Betts. F., Posner, A. S.: Stabilization of amorphous calcium phosphate by Mg and ATP. Calc. Tiss. Res. *23*, 245 (1977)

120. Irving, J. T.: Theories of mineralization of bone. Clin. Orthop. *97*, 225 (1973)

121. Rabinowitz, J. L., Tavares, C. J., Lipson, R., Peterson, P.: Lipid components and in vitro mineralization of some invertebrate cartilages. Biol. Bull. *150*, 69 (1976)

122. Vogel, J. J., Boyan-Salyers, B. D.: Acidic lipids associated with the local mechanism of calcification. A review. Clin. Orthop. *118*, 230 (1976)

123. Wuthier, R. E.: The role of phospholipids in biological calcification. Clin. Orthop. *90*, 191 (1973)

124. Irving, J. T.: The sudanophilic material in the early stages of calcification. Arch. oral Biol. *8*, 735 (1963)

125. Dirksen, T. R., Marinetti, G. V.: Lipids of bovine enamel and dentin and human bone. Calc. Tiss. Res. *6*, 1 (1970)

126. Odutuga, A. A., Prout, R. E. S.: Lipid analysis of human enamel and dentine. Arch. oral Biol. *19*, 729 (1974)

127. Shapiro, I. M.: The phospholipids of mineralized tissues. I. Mammalian compact bone. Calc. Tiss. Res. *5*, 21 (1970)

128. Ennever, J., Vogel, J. J., Levy, B. M.: Lipid and bone matrix calcification in vitro. Proc. Soc. Exp. Biol. (N. Y.) *145*, 1386 (1974)

129. Ennever, J., Vogel, J. J., Benson, L. A.: Lipid and calculus matrix calcification in vitro. J. Dent. Res. *52*, 1056 (1973)

130. Odutuga, A. A., Prout, R. E. S., Hoare, R. J.: Hydroxyapatite precipitation in vitro by lipids extraced from mammalian hard and soft tissues. Arch. oral Biol. *20*, 311 (1975)

131. Shapiro, I. M.: The association of phospholipids with inorganic bone. Calc. Tiss. Res. *5*, 13 (1970)

132. Wuthier, R. E.: The role of phospholipids in biological calcification. Distribution of phospholipase activity in calcifying epiphyseal cartilage. Clin. Orthop. Rel. Res. *90*, 191 (1973)

133. Wuthier, R. E.: Lipid composition of isolated epiphyseal cartilage cells, membranes and matrix vesicles. Biochim. Biophys. Acta *409*, 128 (1975)

134. Majeska, R. J., Wuthier, R. E.: Localization of matrix vesicle phospholipids by labeling with trinitrobenzenesulfonate (TNBS). Intern. Assn. Dent. Res., Preprinted Abstracts. 54th General Session, Abstr. No. 205 (1976)

135. Wuthier, R. E., Gore, S. T.: Partition of inorganic ions and phospholipids in isolated cell, membrane and matrix vesicle fractions: Evidence for $Ca-P_i$-acidic phospholipid complexes. Calc. Tiss. Res. *24*, 163 (1977)

136. Hasiak, R. J., Vadehra, D. V., Baker, R. C.: Lipid composition of the egg exteriors of the chicken *Gallus gallus*. Comp. Biochem. Physiol. *37*, 429 (1970)

137. Hasiak, R. J., Vadehra, D. V., Baker, R. C.: Fatty acid composition of the egg exterior structure of Gallus gallus. Comp. Biochem. Physiol. *35*, 761 (1970)

138. Yabusaki, K. K., Wells, M.: Binding of Calcium to phosphatidyl cholines as determined by proton magnetic resonance and infrared spectroscopy. Biochemistry *14*, 162 (1975)

139. Sutherland, E. W., Oye, I., Butcher, R. W.: The action of epinephrine and the role of the adenyl cyclase system in hormone action. Recent Progr. Hormone Res. *21*, 623 (1965)

140. Vaes, G.: Parathyroid hormone-like action of N^6-2'-O-di-butyryladenosine 3',5'(cyclic)-monophosphate on bone explants in tissue culture. Nature *219*, 939 (1968)

141. Chase, L. R., Aurbach, G. D.: Effect of parathyroid hormone on the concentration of 3', 5'-AMP in bone. Clin. Res. *17*, 380 (1969)

142. Chase, L. R., Melson, G. L., Aurbach, G. D.: Defective excretion of 3'.5'-AMP in response to parathyroid hormone. J. clin. Invest. *48*, 1834 (1969)

143. Herrmann-Erlee, M. P. M.: A parathyroid hormone-like action of dibutyryl cyclic adenosine-3',5'monophosphate on the explanted embryonic mouse radius. Calc. Tiss. Res. *4*, (Suppl.) 70 (1970)

144. Murad, F., Brewer, H. B., Vaughn, M.: Effect of thyreocalcitonin on adenosine 3',5'-cyclic phosphate formation by rat kidney and bone. Proc. Nat. Acad. Sci. (Wash.) *65*, 446 (1970)

145. Peck, W. A., Carpenter, J., Messinger, K., De Bra, D.: Cyclic 3',5'-adenosine monophosphate in isolated bone cells: Response to low concentrations of parathyroid hormone. Endocrinology *92*, 692 (1973)

146. Rodan, S. G., Rodan, G. A.: The effect of parathyroid hormone and thyreocalcitonin on the accumulation of cyclic adenosine 3',5'-monophosphate in freshly isolated bone cells. J. Biol. Chem. *249*, 3068 (1974)

147. Davidovitch, Z., Montgomery, P. C., Eckerdal, O., Gustafson, G. T.: Cellular Localization of Cyclic AMP in Periodontal Tissues during Experimental Tooth Movement in Cats. Calc. Tiss. Res. *19*, 317 (1976)

148. Walsh, D. A., Perkins, J. P., Krebs, E. G.: An adenosine 3',5'-monophosphate dependent protein kinase from rabbit skeletal muscle. J. Biol. Chem. *243*, 3763 (1969)

149. Rasmussen, H., Goodman, D. B. P., Tenenhouse, A.: The role of cyclic AMP and calcium in cell activation. CRC Critical Res. Biochem. *1*, 95 (1972)

150. Lichtenstein, M. L., De Bernardo, R.: The immediate allergie response: In vitro action of cyclic AMP-active and other drugs on the two stages of histamine release. J. Immunol. *107*, 1131 (1971)

151. Kalinger, M., Orange, R. P., Austen, K. F.: Immunological release of histamine and slow reacting substance of anaphylaxis from human lung. J. exp. Med. *136*, 556 (1972)

152. Zurier, R. B., Hoffstein, S., Weissman, G.: Mechanisms of lysosomal enzyme release from human leukocytes. J. Cell. Biol. *58*, 27 (1973)

153. Chase, L. R., Aurbach, G. D.: The effect of parathyroid hormone on the concentration of adenosine 3',5'-monophosphate in skeletal tissue in vitro. J. Biol. Chem. *245*, 1520 (1970)

154. Smith, D. M., Johnston, C. C.: Cyclic 3',5'-adenosine monophosphate levels in separated bone cells. Endocrinology *96*, 1261 (1975)

155. Nagata, N., Sasaki, M., Kimura, N., Nakane, K.: Effects of porcine calcitonin on the metabolism of calcium and cyclic AMP in rat skeletal tissue in vivo. Endocrinology *97*, 527 (1975)

156. Nagata, N., Kimura, N., Sasaki, M., Nakane, K., Tanaka, Y.: Localization of cell groups sensitive to parathyroid hormone and calcitonin in rat skeletal tissue. Biochim. Biophys. Acta *421*, 218 (1976)

157. George, W. J., Polson, J. B., O'Toole, A. G., Goldberg, N. D.: Elevation of guanosine 3',5'-cyclic phosphate in rat heart after perfusion with acetylcholine. Proc. Nat. Acad. Sci. (Wash.) *66*, 398 (1970)

158. Goldberg, N. O., O'Dea, R. F., Haddox, M. K.: Cyclic GMP. In: Advances in cyclic nucleotide research, Vol. 3, pp. 155. Greengard, P., Robison, G. A. (eds.). New York: Raven Press 1973

159. Davidovitch, Z., Montgomery, P. C., Shanfeld, J. L.: Cellular Localization and Concentration of Bone Cyclic Nucleotides in Response to Acute PTE Administration. Calc. Tiss. Res. *24*, 81 (1977)

160. Solomons, C. C.: Peptides obtained from partial hydrolysis of decalcified human dentin collagen. Nature *185*, 101 (1960)

161. Leaver, A. G., Shuttleworth, C. A.: Fractionation of the acid soluble nitrogen of bone and dentine. Arch. oral Biol. *12*, 947 (1967)

162. Leaver, A. G., Eastoe, J. E., Hartles, R. L.: Citrate in mineralized tissues. II. The isolation from human dentine of a complex containing citric acid and a peptide. Arch. oral Biol. *2*, 120 (1960)

163. Leaver, A. G.: The nature and possible significance of peptides isolated from bone and dentine. Calc. Tiss. Res. *2*, (Suppl.) 85 (1968)

164. Krampitz, G., Erben, H. K., Kriesten, K.: On the amino acid composition and structure of egg-shells. Biomineralisation *4*, 88 (1972)

165. Jongebloed, W. L., Van den Berg, P. J., Arends, J.: The dissolution of single crystals of hydroxyapatite in citric and lactic acids. Calc. Tiss. Res. *15*, 1 (1974)

166. Schumacher, G. H., Schmidt, H., Richter, W.: Anatomie und Biochemie der Zähne. pp. 339, Stuttgart: Gustav Fischer Verlag 1976

167. Eastoe, J. E.: Recent studies on the organic matrices of bone and teeth. In: First European Symposium on Bone and Teeth, p. 269. Blackwood, H. J. J. (ed.). Oxford: Pergamon Press 1964

168. Kelly, P. G., Olover, P. T. P., Pautard, F. G. E.: The shell of *Lingula unguis*. Second European Symp. Bones and Teeth, p. 337. Richelle, L. J., Dallemagne, M. J. (eds.). L'Université de Liège 1965

169. Travis, D. F., Francois, C. J., Bonar, L. C., Glimcher, M. J.: Comparative studies of the organic matrices of invertebrate mineralized tissues. J. Ultrastruc. Res. *18*, 519 (1967)

170. Eastoe, J. E.: Chemical Aspects of the Matrix Concept in Calcified Tissue Organization. Calc. Tiss. Res. *2*, 1 (1968)

171. Bonucci, E.: The relationship between mineral substance and organic structures during calcification. Calc. Tiss. Res. Suppl. *24*, Abstr. No 16 (1977)

172. Strates, B., Neuman, W. F., Levins, G. J.: Precipitation of near-neutral solutions of calcium and phosphate. J. Phys. Chem. *61*, 279 (1957)

173. Santanam, M. S.: Calcification of collagen. J. Mol. Biol. *1*, 65 (1959)

174. Bachra, B. N., Sobel, A. E., Stanford, J. W.: Calcification, XXIV. Mineralization of collagen and other fibers. Arch. Biochem. *84*, 79 (1959)
175. Bachra, B. N., Sobez, A. E.: Calcification XXV. Mineralization of reconstituted collagen. Arch. Biochem. *85*, 9 (1959)
176. Glimcher, M. J., Hodge, A. J., Schmitt, F. O.: Macromolecular aggregation states in relation to mineralization. The collagen-hydroxyapatite system as studied in vitro. Proc. Nat. Acad. Sci. (Wash.) *43*, 860 (1957)
177. Glimcher, M. J.: Organic matrices in mineralization. In: Calcification in Biological Systems, pp. 421. Sognaes, R. F. (ed.). Washington, D. C.: Amer. Ass. Adv. Sci. 1960
178. Neuman, W. F., Neuman, M. W.: The chemical dynamics of bone mineral, pp. 169, Chicago. Univ. of Chicago Press 1958
179. Glimcher, M. J., Bonar, L. C., Daniel, E. J.: The molecular structure of the protein matrix of bovine dental enamel. J. Mol. Biol. *3*, 541 (1961)
180. Taves, D. R., Neuman, W. F.: Factors controlling calcification in vitro: Fluoride and magnesium. Arch. Biochem. *108*, 390 (1964)
181. Strates, B., Neuman, F.: On the mechanisms of calcification. Proc. Soc. Exp. Biol. Med. *97*, 688 (1958)
182. Fleisch, H., Neuman, F.: Quantitative aspects of nucleation in calcium phosphate precipitation. J. Amer. Chem. Soc. *82*, 996 (1960)
183. Solomons, C. C., Irving, J. T., Neuman, W. F.: Calcification of the dentin matrix. In: Calcification in biological systems, pp. 203. Sognnaes, R. F. (ed.). Washington: Publ. No. 14, Amer. Ass. Adv. Sci. 1960
184. Schiffman, E., Martin, G. R., Miller, E. J.: Matrices that calcify. In: Biological calcification, cellular and molecular aspects, pp. 27. Schraer, H., (ed.). Amsterdam: North-Holland Publishing Co. 1970
185. Embery, G.: A sulphated Glycopeptide in human supragingival calculus extract. Calc. Tiss. Res. *23*, 13 (1977)
186. Young, S. D.: Collagen in the autoclave-soluble proteins of scleractinian corals (Cnidaria). Comp. Biochem. Physiol. *50B*, 105 (1975)
187. Traub, W., Piez, K. A.: The chemistry and structure of collagen. Adv. Prot. Chem. *25*, 243 (1971)
188. Traub, W.: Some stereochemical implications of the molecular conformation of collagen. Israel J. Chem. *12*, 435 (1974)
189. Boedtker, H., Doty, P.: The native and denatured states of soluble collagen. J. Amer. Chem. Soc. *78*, 4267 (1956)
190. Piez, K. A.: Soluble collagen and the components resulting from its denaturation. In: Treatise on collagen. Vol. I, chemistry of collagen. Ramachandran, G. N. (ed.). New York: Academic Press: 1967
191. Simpson, J. W.: Protein in teeth. In: Dental biochemistry, pp. 24. Lazarri, E. P. (ed.). Philadelphia: Lea & Febiger 1976
192. Fietzek, P., Rexrodt, W.: The covalent structure of collagen. The amino-acid sequence of α2-CH4 of calfskin collagen. Eur. J. Biochem. *59*, 113 (1975)
193. Miller, E. J., Matukas, V. J.: Biosynthesis of collagen. Fed. Proc. *33*, 1199 (1974)
194. Lewis, M. S., Piez, K. A.: Sedimentation-equilibrium studies of the molecular weight of single·and double chains from ratskin collagen. Biochemistry *3*, 1126 (1964)
195. Dehm, P., Prockop, D. J.: Synthesis and extrusion of collagen by freshly isolated cells from chick embryo tendon. Fed. Proc. *30*, 1196 (1971)
196. Piez, K. A., Eigner, E. A., Lewis, M. S.: The chromatographic separation and amino acid composition of the subunits of several collagens. Biochemistry *2*, 58 (1963)
197. Miller, E. J., Martin, G. R., Piez, K. A., Powers, M. J.: Characterization of chick bone collagen and compositional changes assiciated with maturation. J. Biol. Chem. *242*, 5481 (1967)
198. Volpin, D., Veis, A.: Isolation and characterization of the cyanogen-bromide peptides from the α1 and α2 chains of acid-soluble bovine skin collagen. Biochemistry *10*, 1751 (1971)
199. Spiro, R. G.: Studies on the renal glomerular basement membrane. Nature of the carbohydrate units and their attachment to the peptide portion. J. Biol. Chem. *242*, 1923 (1967)

200. Spiro, R. G.: The structure of the disaccharide unit of the reval glomerular basement membrane. J. Biol. Chem. *242,* 4813 (1967)

201. Spiro, R. G.: The carbohydrate of collagens. In: Chemistry and molecular biology of the intercellular matrix, Vol. 1, pp. 195. Balazs, E. A. (ed.). London, New York: Academic Press 1970

202. Veis, A., Bhatnagar, R. S.: The microfibrillar structure of collagen and the placement of intermolecular covalent crosslinkages. In: Chemistry and molecular biology of the intercellular matrix, Vol. 1, pp. 279. Balazs, E. A. (ed.). London, New York: Academic Press 1970

203. Mechanic, G.: Stable intermolecular cross-links in collagen after reduction. In: Chemistry and molecular biology of the intercellular matrix, Vol. 1, pp. 305. Balazs, E. A. (ed.). London, New York: Academic Press 1970

204. Eastoe, J. E., Martens, P., Thomas, N. R.: The amino acid composition of human hard tissue collagens in osteogenesis imperfecta and dentinogenesis imperfecta. Calc. Tiss. Res. *12,* 91 (1973)

205. Hayashi, Y., Igarashi, M.: The change of lysyl residues in the bone collagen in experimental osteoporosis. Calc. Tiss. Res. *13,* 163 (1973)

206. Furukawa, T., Iwata, H., Nakagawa, M., Kuboki, Y., Sasaki, S.: Changes of collagen molecule in human osteoarthritic bone lesions. Calc. Tiss. Res. *23,* 197 (1977)

207. Kuftinec, M. M., Miller, S. A.: Bone growth in the neonatal rat I. Biochemical aspects of bone protein synthesis. Calc. Tiss. Res. *11,* 105 (1973)

208. Wergedal, J. E.: Enzymes of protein and phosphate catabolism in rat bone. Enzyme property in normal rat. Calc. Tiss. Res. *3,* 55 (1969)

209. Kuftinec, M. M., Miller, S. A.: Bone growth in the neonatal rat I. Biochemical aspects of bone protein synthesis. Calc. Tiss. Res. *11,* 105 (1973)

210. Herring, G. M.: The organic matrix of bone. In: The biochemistry and physiology of bone. Vol. 1, pp. 127. Bourne (ed.). New York: Academic Press 1972

211. Leaver, A. G., Shuttleworth, C. A.: Studies on the peptides, free amino acids and certain related compounds from ox bone. Arch. oral Biol. *13,* 509 (1968)

212. Andrews, A. T. de B., Herring, G. M., Kent, P. W.: The periodate oxidation of bovine sialoprotein and some observations on its structure. Biochem. J. *111,* 621 (1969)

213. Andrews, A. T. de B., Herring, G. M., Kent, P. W.: Some studies on the composition of bovine cortical-bone sialoprotein. Biochem. J. *104,* 705 (1967)

214. Wyckoff, R. W. G.: The biochemistry of animal fossils, pp. 71, Bristol: Scientechnica Ltd. 1972

215. Herring, G. M., Ashton, B. A., Chipperfield, A. R.: The isolation of soluble proteins, glycoproteins, and proteoglycans from bone. Prep. Biochem. *4,* 179 (1974)

216. Herring, G. M.: Methods for the study of the glycoproteins and proteoglycans of bone using bacterial collagenase. Determination of bone sialoprotein and chondroitin sulphate. Calc. Tiss. Res. *24,* 29 (1977)

217. Vetter, G.: personal communication (1978)

218. Lam, K. W., Li, O., Li, C. Y., Yam, L. T.: Biochemical Properties of human prostatic acid phosphatase. Clin. Chem. *19,* 483 (1973)

219. Vatassery, G. T., Singer, L., Armstrong, W. D.: Hydrolysis of pyrophosphate and ester phosphates by bone extracts. Calc. Tiss. Res. *5,* 180 (1970)

220. Toverud, S. U., Hammarström, L. E., Kristoffersen, U. M.: Quantitative studies on acid phosphatase in developing rat bones and teeth during hypervitaminosis D. Arch. oral Biol. *20,* 175 (1975)

221. Lieberherr, M., Vreven, J., Vaes, G.: The acid and alkaline phosphatases, inorganic pyrophosphatases and phosphoprotein phosphatase of bone. I. Characterization and assay. Biochim. Biophys. Acta *293,* 160 (1973)

222. Anderson, T. R., Toverud, S. U.: Chromatographic separation of two acid phosphatases from rat bone. Calc. Tiss. Res. *24,* 187 (1977)

223. Tolna, S.: Effects of parathyroid hormone on bone acid hydrolases in tissue culture. Canad. J. Physiol. Pharmacol. *46,* 261 (1968)

224. Vaes, G.: On the mechanism of bone resorption: the action of PTH on the excretion and synthesis of lysosomal enzymes and the extracellular release of acid bone cells. J. Cell. Biol. *39*, 676 (1968)

225. Horman, A. W., Henry, H.: 1,25-dihydroxy cholecalciferol, a hormonally active form of vitamin D_3. Recent Prog. Horm. Res. *30*, 431 (1974)

226. Morgan, D. B., Monod, A., Russell, R. G. G., Fleisch, H.: Influence of dichlormethylene diphosphonate (Cl_2MDP) and calcitonin on bone resorption, lactate production and phosphate and pyrophosphatase content of mouse calvaria treated with parathyroid hormone in vitro. Calc. Tiss. Res. *13*, 287 (1973)

227. Fell, H. B., Dingle, G. T.: Studies on the mode of action of excess vitamin A. 6. Lysosomal protease and the degradation of cartilage matrix. Biochem. J. *87*, 403 (1963)

228. Lieberherr, M., Pezant, E., Garabedian, M., Balsan, S.: Phosphatase content of rat calveria after in vivo administration of vitamin D_3 metabolites. Calc. Tiss. Res. *83*, 235 (1977)

229. Linde, A., Granström, G., Magnusson, B. C.: Ca^{2+}-ATPase in hard tissue forming cells. Calc. Tiss. Res. *81*, 108 (1976)

230. Owen, M., Trifitt, J. T., Mellick, R. A.: Albumine in bone. In: Hard tissue growth, repair and remineralization. Ciba Fnd. Symp., no. 11, pp. 263. Amsterdam: Assoc. Sci. Publishers 1973

231. Owen, M., Triffitt, J. T.: Plasma glycoproteins and bone. In: Calcium parathyroid hormone and the calcitonins, pp. 316. Talmage, R. V., Munson, P. L. (eds.). Amsterdam: Excerpta Medica 1972

232. Wilson, J. M., Ashton, B., Triffitt, J. T.: Interaction of a Component of bone organic matrix with the mineral phase. Calc. Tiss. Res. *22*, Suppl. 458 (1977)

233. Ashton, B. A., Triffitt, J. T., Herring, G. M.: Isolation and partial characterization of a glycoprotein from bovine cortical bone. Europ. J. Biochem. *45*, 525 (1974)

234. Triffitt, J. T., Owen, M.: Studies of bone matrix glycoproteins. Incorporation of $1-^{14}C$ glucosamine and plasma [^{14}C] glycoprotein into rabbit cortical bone. Biochem. J. *136*, 125 (1973)

235. Miller, L. L., John, D. W.: Nutritional, hormonal, and temporal factors regulating net plasma protein biosynthesis in the isolated perfused rat liver. In: Plasma protein metabolism, pp. 207. Rothschild, M. A., Waldmann, T. (eds.). New York, London: Academic Press 1970

236. Triffitt, J. T., Gebauer, U., Owen, M. E.: Synthesis by the liver of a glycoprotein which is concentrated in bone matrix. Calc. Tiss. Res. *21*, Suppl. 437 (1976)

237. Triffitt, J. T., Owen, M.: Preliminary studies on the binding of plasma albumine to bone tissue. Calc. Tiss. Res. *23*, 103 (1977)

238. Ashton, B. A., Triffitt, J. T., Höhling, H. J.: Plasma proteins present in human cortical bone: Enrichment of the α_2 HS-glycoproteins. Calc. Tiss. Res. *22*, 27 (1976)

239. Owen, M. E., Triffitt, J. T.: Extravascular albumin in bone tissue. J. Physiol. *257*, 293 (1976)

240. Owen, M., Howlett, C. R., Triffitt, J. T.: Movement of ^{125}J Albumin and ^{125}J Polyvinyl-pyrrolidone through bone tissue fluid. Calc. Tiss. Res. *23*, 103 (1977)

241. Vittur, F., Zanet, M., Stagni, N., De Bernard: Role of proteoglycans on calcification. Calc. Tiss. Res. *24*, Abstr. No. 99 (1977)

242. Dickson, I.: The composition and antigenicity of sheep cortical bone matrix proteins. Calc. Tiss. Res. *16*, 321 (1974)

243. Carmichael, D. J., Veis, A., Wang, E. T.: Dentin matrix collagen: evidence for a covalently linked phosphoprotein attachment. Calc. Tiss. Res. *7*, 331 (1971)

244. Herring, G. M.: Studies on the protein-bound chondroitin sulphate of bovine cortical bone. Biochem. J. *107*, 41 (1968)

245. Herring, G. M.: Mucosubstances of cortical bone. In: First European Bone and Tooth Symposium, pp. 263. Blackwood, H. J. J. (ed.). Oxford: Pergamon Press 1964

246. Andrews, A. T. de B., Herring, G. M., Kent, P. W.: Some studies on the composition of bovine cortical-bone sialoprotein. Biochem. J. *104*, 705 (1967)

247. Andrews, A. T. de B., Herring, G. M., Kent, P. W.: The periodate oxidation of bovine sialoprotein and some observations on its structure. Biochem. J. *111*, 621 (1969)

248. Bonucci, E.: The locus of initial calcification in cartilage and bone. Clin. Orthop. *78*, 108 (1971)

249. Bowness, J. C.: Present concepts of the role of ground substance in calcification. Clin. Orthop. *59*, 233 (1968)

250. Campo, R. D., Dziewiatkowski, D. D.: Turnover of the organic matrix of cartilage and bone as visualizes by autoradiography. J. Cell. Biol. *18*, 19 (1963)

251. Herring, G. M.: A review of recent advances in the chemistry of calcifying cartilage and bone matrix. Calc. Tiss. Res. *4*, 17 (1970)

252. Antonopoulos, C. A., Engfeldt, B., Gardell, S., Hjertquist, S. Q., Solheim, K.: Isolation and identification of the glycosaminoglycans from fractured callus. Biochim. Biophys. Acta *101*, 150 (1965)

253. Salvo, J. D., Schubert, M.: Specific interaction of some cartilage proteinpolysaccharides with freshly precipitating calcium phosphate. J. Bio. Chem. *242*, 705 (1967)

254. Woodward, C., Davidson, E. A.: Structure function relationship of protein polysaccharide complexes: Specific ionbinding properties. Proc. Nat. Acad. Sci. (Wash.) *60*, 201 (1968)

255. Campo, R. D.: Protein-polysaccharides of cartilage and bone in health and disease. Clin. Orthop. *68*, 182 (1970)

256. Mourao, P. A. S., Ròzenfeld, S., Laredo, J., Dietrich, C. P.: The distribution of chondroitin sulfates in articular and growth cartilages of human bones. Biochim. Biophys. Acta *428*, 19 (1976)

257. Oohira, A., Tamaki, K., Terashima, Y., Chiba, A., Nogami, H.: Glycosaminoglycans in congenital pseudoarthrosis. Calc. Tiss. Res. *23*, 271 (1977)

258. Cuervo, L. A., Pita, J. C., Howell, D. S.: Inhibition of calcium phosphate mineral growth by proteoglycan aggregate fractions in a synthetic lymph. Calc. Tiss. Res. *13*, 1 (1973)

259. Larsson, S. E., Ray, R. D., Kuettner, K. E.: Microchemical studies an acid glycosaminoglycans of the epiphyseal zones during endochondral calcification. Calc. Tiss. Res. *13*, 271 (1973)

260. Larsson, S. E.: The metabolic heterogeneity of glycosaminoglycans of the different zones of the epiphyseal growth plate and the effect of ethane-1-hydroxy-1,1-diphosphonate (EHDP) upon glycosaminoglycan synthesis in vivo. Calc. Tiss. Res. *21*, 67 (1976)

261. Lindenbaum, A., Kuettner, K. E.: Mucopolysaccharides and mucoproteins of calf scapula. Calc. Tiss. Res. *1*, 153 (1967)

262. Hjertquist, S. O.: The glycosaminoglycans of the epiphyseal plates in normal and redutic dogs. Acta Soc. Med. Upsalien. *69*, 83 (1964)

263. Larsson, S. E., Lempberg, R. K.: The glycosaminoglycans of the different layers of bovine articular cartilage in relation to age. II. Incorporation of ^{35}S-sulphate in vitro into different fractions of chondroitin sulfate. Calc. Tiss. Res. *15*, 253 (1974)

264. Larsson, S. E., Kuettner, K. E.: Microchemical studies of acid glycosaminoglycans from isolated chondrocytes in suspension. Calc. Tiss. Res. *14*, 49 (1974)

265. Pedrini-Mille, A., Pedrini, V., Ponsetti, I. V.: Glycosaminoglycans of iliac crest cartilage in spondyloepiphyseal displasia congenita. Calc. Tiss. Res. *16*, 183 (1974)

266. Burgess, R. C., Maclaren, C.: Proteins in developing bovine enamel. In: Tooth enamel: Its composition, properties, and fundamental structure, pp. 74. Stack, M. V., Fearnhead, R. W. (eds.). Bristol: Wright 1965

267. Eastoe, J. E.: The amino acid composition of proteins from oral tissues: II. The matrix protein in dentine and enamel from developing human deciduous teeth. Arch. oral Biol. *8*, 633 (1963)

268. Eggert, F. M., Allen, G. R., Burgess, R. C.: Purification and partial characterization of proteins from developing bovine dental enamel. Biochem. J. *131*, 471 (1973)

269. Elwood, W. K., Apostolopoulos, A. X.: Analysis of developing enamel of the rat: II. Electrophoretic and amino acid studies. Calc. Tiss. Res. *17*, 327 (1975)

270. Glimcher, M. J., Mechanic, G. L., Friberg, U. A.: The amino acid composition of the organic matrix and the neutral soluble and acid soluble components of embryonic bovine enamel. Biochem. J. *93*, 198 (1964)

271. Levine, P. T., Seyer, J. M., Huddleston, J., Glimcher, M. J.: The comparative biochemistry of the organic matrix proteins of the developing enamel: I. Amino acid composition. Arch. oral Biol. *12*, 407 (1967)

272. Bonar, L. C., Glimcher, M. J., Mechanic, G. L.: The molecular structure of the neutral-soluble proteins of embryonic bovine enamel in the solid state. J. Ultrastruct. Res. *13*, 308 (1965)

273. Bonar, L. C., Mechanic, G. L., Glimcher, M. J.: Optical rotatory dispersion studies of the neutral soluble proteins of embryonic bovine enamel. J. Ultrastruct. Res. *13*, 296 (1965)

274. Glimcher, M. J., Bonar, L. C., Daniel, E. J.: The molecular structure of the protein matrix of bovine dental enamel. J. Mol. Biol. *3*, 541 (1961)

275. Fincham, A. G.: Electrophoretic and sephadex gel filtration studies of bone foetal enamel matrix at acid pH. Calc. Tiss. Res. *2*, 353 (1968)

276. Katz, E. P., Mechanic, G. K., Glimcher, M. J.: The ultracentrifugal and free zone electrophoretic characterization of the neutral soluble proteins of embryonic bovine enamel. Biochim. Biophys. Acta *107*, 471 (1965)

277. Katz, E. P., Seyer, J. M., Levine, P. T., Glimcher, M. J.: Ultracentrifugal and electrophoretic characterization of proteins soluble at neutral pH. Arch. oral Biol. *14*, 533 (1969)

278. Mechanic, G. L.: The multicomponent re-equilibrating protein system of bovine embryonic enamelin (dental enamel protein). In: Tooth enamel II. Its composition, properties and fundamental structure, pp. 88. Fearnhead, R. W., Stack, M. V. (eds.). Bristol: Wright 1971

279. Mechanic, G. L., Katz, E. P., Glimcher, M. J.: The sephadex gel filtration characteristics of the neutral soluble proteins of embryonic bovine enamel. Biochim. Biophys. Acta *133*, 97 (1967)

280. Seyer, J. M., Glimcher, M. J.: Evidence for the presence of numerous protein components in immature bovine dental enamel. Calc. Tiss. Res. *24*, 253 (1977)

281. Fukae, M.: Biochemical studies of enamel formation. J. Biochem. *37*, 295 (1970)

282. Fukae, M., Shimizu, M.: Studies on the proteins of developing bovine enamel. Arch. oral Biol. *19*, 381 (1974)

283. Fukae, M., Shimokawa, H., Iwakura, M., Suda, T., Sasaki, S.: Studies of the biosynthesis of enamel proteins in the rat by use of ^{3}H-proline (abstract). J. Dent. Res. *51*, 1298 (1972)

284. Guenther, H. L., Croissant, R. C., Schonfeld, S., Slavkin, H. C.: Enamel proteins: Identification of epithelial-specific differentiation products. In: Extracellular matrix influences on gene expression, pp. 387. Slavkin, H. C., Greulich, R. C. (eds.). New York: Academic Press 1975

285. Shimokawa, H., Sasaki, S.: Biosynthesis of enamel protein in vitro (abstract). J. Dent. Res. *54A*, 107 (1975)

286. Deakins, M.: Changes in the ash, water, and organic content of pig enamel during calcification. J. Dent. Res. *21*, 429 (1942)

287. Glimcher, M. J., Friberg, U. A., Levine, P. T.: The isolation and amino acid composition of the enamel proteins of erupted bovine teeth. Biochem. J. *93*, 202 (1964)

288. Stack, M. V.: Chemical organization of the organic matrix of enamel. In: Structural and chemical organization of teeth, Vol. 2, pp. 317. Miles, A. E. W. (ed.). New York: Academic Press 1967

289. Weinmann, J. P., Wessinger, J. D., Reed, G.: Correlation of chemical and histological investigations on developing enamel. J. Dent. Res. *21*, 171 (1942)

290. Glimcher, M. J., Brickley-Parsons, D., Levine, P. T.: Studies of enamel proteins during maturation. Calc. Tiss. Res. *24*, 259 (1977)

291. Robinson, C., Lowe, N. R., Weatherell, J. A.: Changes in amino-acid composition of developing rat in incisor enamel. Calc. Tiss. Res. *23*, 19 (1977)

292. Schonfeld, S. D., Slavkin, H. C.: Demonstration of enamel matrix proteins on root-analogie surfaces of rabbit permanent incisor teeth. Calc. Tiss. Res. *24*, 223 (1977)

293. Guenther, H. L., Croissant, R. C., Schonfeld, S. E., Slavkin, H. C.: Identification of four extracellular matrix enamel proteins during embryonic rabbit tooth-organ development. Biochem. J. *163*, 591 (1977)

294. Eastoe, J. E.: Dental Enamel. In: Comprehensive biochemistry, Vol. 26, pp. 785. Florkin, M., Stotz, E. (eds.). New York: Elsevier 1971

295. Glimcher, M. J., Krane, S. M.: The identification of serine phosphate in enamel proteins. Biochim. Biophys. Acta *90*, 477 (1964)

296. Krane, S. M., Stone, M. J., Glimcher, M. J.: The presence of protein phosphokinase in connective tissues and the phosphorylation of enamel protein in vitro. Biochim. Biophys. Acta *97*, 77 (1965)

297. Patterson, C. M. Kruger, B. J., Daley, T. J.: Lipid and protein histochemistry of enamel-effects of fluoride. Calc. Tiss. Res. *24*, 119 (1977)

298. Elwood, W. K., Apostolopoulos, A. X.: Analysis of developing enamel of the rat. III. Carbohydrate, DEAE-sephadex, and immunological studies. Calc. Tiss. Res. *17*, 337 (1975)

299. Veis, A., Schleuter, R. J.: The macromolecular organization of dentine matrix collagen. I. Characterization of dentine collagen. Biochemistry *3*, 1650 (1964)

300. Araya, S., Saito, S., Nakanishi, A., Kawanishi, Y.: Physiology, Soluble collagen in Bone. Nature *192*, 758 (1961)

301. Saito, S.: Biochemical investigation of the peridontal membrane. (1) On the electronmicroscopical observation of protein component. Bull. Tokyo med. dent. Univ. 7, 385 (1960)

302. Veis, A.: Intact collagen. In: Treatise on collagen. Vol. I. pp. 367. Ramachandran, G. N. (ed.). New York: Academic Press 1967

303. Ranta, H., Bailey, A. J.: Age-related changes in the cross-linking of dentine collagen. Calc. Tiss. Res. *24*, Suppl. abstract No. 81 (1977)

304. Carmichael, D. J. Dodd, K. M., Nawrot, C. F.: Studies on matrix proteins of normal and lathyritic rat bone and dentine. Calc. Tiss. Res. *14*, 177 (1974)

305. Leaver, A. G., Thomas, Holbrook, I. B.: Glycoproteins of Human Dentine. Calc. Tiss. Res. *22*, 347 (1977)

306. Veis, A., Perry, A.: The phosphoprotein of dentin matrix. Biochemistry *6*, 2409 (1967)

307. Butler, W. T., Finch, J. E., De Steno, C. V.: Chemical character of proteins in rat incisors. Biochim. Biophys. Acta *257*, 167 (1972)

308. Jontell, M., Linde, A.: Phosphoprotein of rat incisor dentine. Calc. Tiss. Res. *22*, 321 (1977)

309. Jones, I. L., Leaver, A. G.: Glycosaminoglycans of human dentine. Calc. Tiss. Res. *16*, 37 (1974)

310. Takuma, S., Yanagisawa, T., Li, W. L.: Ultrastructural and microanalytical aspects of developing osteodentin in rat incisors. Calc. Tiss. Res. *24*, 215 (1977)

311. Ellingson, J. S., Smith, M., Larson, L. R.: Phospholipid composition and fatty acid profiles of the phospholipids of bovine predentin. Calc. Tiss. Res. *24*, 127 (1977)

312. Alam, S. Q., Alvarez, J. C., Harris, R. S.: Effects of nutrition on the composition of tooth lipids and fatty acids in rats. III. Effects of feeding different oils and fats on caries and on fatty acid composition of teeth. J. Dent. Res. *52*, 229 (1973)

313. Prout, R. E. S., Atkin, (nee Shutt), E. R.: Effect of diet deficient in essential fatty acids on fatty acid composition of enamel and dentine of the rat. Arch. oral Biol. *18*, 583 (1973)

314. Prout, R. E. S., Odutuga, A. A.: In vivo incorporation of $[1-{}^{14}C]$-linoleic acid into the lipids of enamel and dentin of normal and essential fatty acid deficient rats. Arch. oral Biol. *19*, 1167 (1974)

315. Odutuga, A. A., Prout, R. E. S.: Fatty acid composition of carious molar enamel and dentin from rats deficient in essential fatty acids. Arch. oral Biol. *20*, 49 (1975)

316. Odutuga, A. A., Prout, R. E. S.: Fatty acid composition of neutral lipids and phospholipids of enamel and dentin from rat incisors and molars. Arch. oral Biol. *18*, 689 (1973)

317. Prout, R. E. S., Odutuga, A. A.: Fatty acid composition of neutral lipids and phospholipids of enamel and dentin from human molars. Arch. oral Biol. *19*, 293 (1974)

318. Birkedal-Hansen, H., Butler, W. T., Taylor, R. E.: Proteins of the periodontinum. Characterization of the insoluble collagens of bovine dental cementum. Calc. Tiss. Res. *23*, 39 (1977)

319. Rodriguez, M. S., Wildermann, M. N.: Amino acid composition of the cementum matrix from human molar teeth. J. Periodont. *43*, 438 (1972)

320. Rodriguez, M. S., Wildermann, M. N.: Amino acid composition of the cementum matrix from human molar teeth. J. Periodont. *43*, 438 (1972)

321. Birkedal-Hansen, H., Butler, W. T., Taylor, R. E.: Isolation and preliminary characterization of bovine cementum collagen. Calc. Tiss. Res. *15*, 325 (1974)

131

322. Miller, E. J.: A review of biochemical studies on the genetic distinct collagens of the skeletal system. Clin. Orthop. *92*, 260 (1973)

323. Epstein, E. H.: [α 1(III)] $_3$ human skin collagen. Release by pepsin digestion and preponderance in fetal life. J. Biol. Chem. *249*, 3225 (1974)

324. Butler, W. T., Birkedal-Hansen, H., Beegle, W. F., Taylor, R. E., Chung, E.: Proteins of periodontum. Identification of collagens with the [α1 (I)$_2$]α^2 and [α1 (III)]$_3$ structures in bovine periodontal ligament. J. Biol. Chem. *250*, 8907 (1975)

325. Christner, P., Robinson, P., Clark, C. C.: A preliminary characterization of human cementum collagen. Calc. Tiss. Res. *23*, 147 (1977)

326. Wells, H. G.: The chemistry of arteriosclerosis. A survey of the problem, p. 323. Cowdry, E. V. (ed.). New York: Macmillan 1933

327. Weissman, G., Weissman, S.: X-ray diffractions studies of human aortic elastin. J. clin. Invest. *39*, 1657 (1960)

328. Urry, D. W.: Neutral sites for calcium ion binding to elastin and collagen: A charge neutralization theory for calcification and its relationship to arteriosclerosis. Proc. Nat. Acad. Sci. (Wash.) *68*, 810 (1971)

329. Urry, D. W.: On the molecular basis for vascular calcification. Perspect. Biol. Med. *18*, 68 (1974)

330. Partridge, S. M.: Elastin. Adv. in Prot. Chem. *17*, 227 (1962)

331. Partridge, S. M.: Isolation and characterization of elastin. In: Chemistry and molecular biology in the intercellular matrix, Vol. 1, pp. 593. Balazs, E. A. (ed.). London, New York: Academic Press 1970

332. Franzblau, C., Faris, B., Lent, R. W., Salcedo, L. L., Smith, B., Jaffe, R., Cromby, G.: Chemistry and biosynthesis of cross-links in elastin. In: Chemistry and molecular biology of the intercellular matrix, Vol. 1, pp. 617. Balazs, E. A. (ed.). London, New York: Academic Press 1970

333. Ross, R., Bornstein, P.: Studies of the components of the elastin fiber. In: Chemistry and molecular biology of the intercellular matrix, Vol. 1, pp. 641. Balazs, E. A. (ed.). London, New York: Academic Press 1970

334. Yu, S. Y., Blumenthal, H. T.: The calcification of elastin fibers. II. Ultramicroscopic characteristics. J. Geront. *18*, 127 (1963)

335. Sobel, A. E., Leibowitz, S., Eilberg, R. G., Lamy, F.: Nucleation by elastin. Nature *211*, 45 (1966)

336. Yu, S. Y., Blumenthal, H. T.: The calcification of elastic tissue. In: The connective tissue, pp. 17. Wagner, B. M., Smith, D. E. (eds.). Baltimore: Williams & Wilkens 1967

337. Seligman, M., Eilberg, R. G., Fishman, L.: Mineralization of elastin extracted from human aortic tissues. Calc. Tiss. Res. *17*, 229 (1975)

338. Eilberg, R. G., Mori, K.: Early stages of in vitro calcification of human aortic tissue. Nature *223*, 518 (1969)

339. Lansing, A. I., Roberts, E., Ramasarma, G. B., Rosenthal, T. B., Alex, M.: Changes with age in amino acid composition of arterial elastin. Proc. Soc. exp. Biol. *74*, 714 (1951)

340. Franzblau, C., Faris, B., Papaioannou, P. L.: Lysinonorleucine. A new amino acid from hydrolyzates of elastin. Biochemistry *8*, 2833 (1969)

341. Schiffmann, E., Martin, G. R., Corcoran, B. A.: The role of matrix in aortic calcification. Arch. Biochem. Biophys. *107*, 284 (1964)

342. Schiffmann, E., Corcoran, B. A., Martin, G. R.: Role of complexed heavy metals initiating mineralization of "elastin" and the precipitation of mineral from solution. Arch. Biochem. Biophys. *115*, 87 (1966)

343. Schiffmann, E., Lavender, D. R., Miller, E. J., Corcoran, R. A.: Amino acids of the nucleation site in mineralizing tissue. Calc. Tiss. Res. *3*, 125 (1969)

344. Eisenstein, R., Ayer, J. P., Papajiannis, S., Hass, G. M., Ellis, H.: Mineral binding by human arterial elastic tissue. Lab. Invest. *13*, 1198 (1964)

345. Molinari-Tosatti, M. P., Gotte, L.: Some features of the binding of calcium ions to elastin. Calc. Tiss. Res. *6*, 329 (1971)

346. Hall, D. A.: The reaction between elastase and elastic tissue. Biochem. J. *59*, 459 (1955)

347. Molinari-Tosatti, M. P., Galzignan, L., Moret, V., Gotte, L.: Some features of the binding of calcium ions to elastin. Calc. Tiss. Res. 2, 88 (1968)

348. Starcher, B. C., Saccomari, G., Urry, D. W.: Coacervation and ion-binding studies on aortic elastin. Biochim. Biophys. Acta 310, 481 (1973)

349. Keller, S., Mandl, I.: Non-polar peptides from elastin. In: Proteins and related subjects, Vol. 22, pp. 127. Peeters, H. (ed). Oxford, New York, Toronto, Braunschweig, Sidney: Pergamon Press 1975

350. Foster, J. A., Bruenger, E., Gray, W. R., Sanberg, L. B.: Isolation and amino acid sequences of tropoelastin peptides. J. Biol. Chem. 248, 2876 (1973)

351. Urry, D. W., Onishi, M.: In: Spectroscopic approaches to biomolecular conformation, pp. 86. Urry, D. W. (ed.). Chicago: American Medical Association Press 1970

352. Geddes, A. J., Parker, K. D., Atkins, E. D. T., Beighton, E.: "Cross-β" conformation in proteins. J. Mol. Biol. 32, 342 (1968)

353. Urry, W. D., Krivacia, J. R., Haider, J.: Calcium ion effects of notable change in elastin conformation by interacting with neutral sites. Biochem. Biophys. Res. Commun. 43, 6 (1971)

354. Urry, D. W., Cummingham, W. D., Onishi, T.: A neutral polypeptide calcium ion complex. Biochim. Biophys. Acta 292, 853 (1973)

355. Starcher, B., Urry, D. W.: Elastin coacervate as a matrix for calcification. Biochem. Biophys. Res. Commun. 53, 210 (1973)

356. Rucker, R. B., Ford, D., Goettlich-Riemann, W., Tom, K.: Additional evidence for the binding of calcium ions to elastin at neutral sites. Calc. Tiss. Res. 14, 317 (1974)

357. Ross, R.: The elastin fiber, a review. J. Histochem. Cytochem. 21, 199 (1973)

358. Starcher, B. C., Cox, B. A., Urry, D. W.: Development of an in vitro system for the calcification of tropoelastin and α-elastin coacervates in serum. Calc. Tiss. Res. 17, 1 (1974)

359. Urry, D. W., Hendrix, C. F., Long, M. M.: Calcification of α-elastin-coacervates: a bulk property of elastin. Calc. Tiss. Res. 21, 57 (1976)

360. Krampitz, G., Greuel, E., Kriesten, K., Hardebeck, H., Engels, J., Koester, U., Helfgen, I., Arbabi, F.: Beiträge zur Molekularbiologie der Eischalenbildung. Berl. Münch. Tierärztl. Wschr. 86, 313 (1973)

361. Wedral, E., Vadehra, D. V., Baker, R. C.: Chemical composition of the cuticle, and the inner and outer shell membranes from eggs of Gallus gallus. Comp. Biochem. Physiol. 47B, 631 (1974)

362. Krampitz, G., Kriesten, K., Faust, R.: Über die Aminosäurenzusammensetzung morphologischer Eischalen-Fraktionen von Ratitae. Biomineralisation 7, 1 (1974)

363. Krampitz, G., Engels, J., Heindl, I., Heinrich, A., Hamm, M., Faust, R.: Biochemische Untersuchungen an Eischalen. Arch. f. Geflügelkde. 38, 197 (1974)

364. Wolken, J. J.: Structure of the hen's egg membranes. Analyt. Rec. 111, 79 (1951)

365. Baker, J. R., Balch, D. A.: A study of the organic material of hen's egg shell. Biochem. J. 82, 352 (1962)

366. Simkiss, K., Taylor, T. G.: Shell formation. In: Physiology and biochemistry of the domestic fowl, Vol. 3, pp. 1331. Bell, D. J., Freeman, B. M., (eds.). London, New York: Academic Press 1971

367. Vadehra, D. V., Nath, K. R., Baker, R. C.: Physiochemical properties of the protein of egg shell membrane. Poult. Sci. 50, 1638 (1971)

368. Balch, D. A., Cooke, R. A.: A study of the composition of hen's egg shell membrane. Ann. Biol. anim. Biochem. Biophys. 10, 13 (1970)

369. Candlish, J. K., Scougall, R. G.: L-5-Hydroxyglycine as a constituent of the shell membrane of the hen's egg. Int. J. Prot. Res. 1, 299 (1969)

370. Simons, P. C. M.: Ultrastructure of the hen egg shell and its physiological interpretation. Agricultural Research Reports 758, Centre for Agricultural Publishing and Documentation, Wageningen 1971

371. Heinrich, A.: Extraktion und Spaltung von Proteinen der Eischalenmembran (Huhn). Dissertation, Universität Bonn 1975

372. Paul-Gardais, A., Picard, J., Hermelin, B.: Turnover of sulfate in glycopeptides from egg shell membranes and hen oviduct sulfated glycoproteins. Biochim. Biophys. Acta *354*, 11 (1974)

373. Siewert, E., Bronsch, K.: Die Eischale und die Zusammensetzung des Eies. In: Handbuch der Tierernährung, Bd. 2. pp. 645. Berlin: Parey 1972

374. Krampitz, G.: Die Eischale. In: Handbuch für Geflügelphysiologie, Kapitel 15. Mehner, A. (ed.). Jena: Gustav Fischer Verlag (in press)

375. Schmidt, W. J.: Goldfärbung der Schalenhautfasern des Vogeleies im Schliff. Z. f. wiss. Mikroskop. *67*, 51 (1965)

376. Simkiss, K., Tyler, C.: A histochemical study of the organic matrix of the hen's egg shell. Quart. J. micr. Sci. *98*, 12 (1967)

377. Bronsch, K., Diamantstein, T.: Beiträge zur Biochemie der Eischalenbildung. 2. Mitt.: Über Mucopolysaccharide der Eischale. Zentralbl. Vet. Med. *2*, 323 (1964)

378. Krampitz, G., Koester, U., Fischer, W.: Vergleichende Untersuchungen der Aminosäuren-Komposition von Vogeleischalen. Biochemische und taxonomische Beziehungen zwischen Hühner-, Kranich- und Gänsevögeln. Z. zool. Systematik u. Evolutionsforsch. *13*, 125 (1975)

379. Krampitz, G., Kriesten, K., Boehme, W.: Untersuchungen über Ultrastruktur und Amino-säurenzusammensetzung der Eischale von *Natrix natrix*. Experientia *29*, 416 (1973)

380. Krampitz, G., Boehme, W., Kriesten, K., Hardebeck, H.: Die Aminosäurenzusammensetzung von Reptilieneischalen in biochemischer und evolutiver Sicht. Z. zool. Systematik u. Evolutionsforsch. *12*, 1 (1974)

381. Krampitz, G., Engels, J.: Beiträge zur Chromatographie von wasserlöslichen Biopolymeren aus der Eischalen-Matrix. Biomineralisation *8*, 21 (1975)

382. Krampitz, G., Engels, J.: Water-soluble proteins from egg-shell matrices. In: Proteins and related subjects, Vol. 22, pp. 327. Peeters, H. (ed.). Oxford, New York, Toronto, Braunschweig, Sidney: Pergamon Press 1975

383. Krampitz, G., Sülz, M., Hardebeck, H.: Die Toxizität von Cadmium beim Huhn. Arch. Geflügelkde. *38*, 86–90 (1974)

384. Hardebeck, H. Sülz, H., Krampitz, G.: Verteilung von Cadmium im Organismus. Arch. Geflügelkde. *38*, 100 (1974)

385. Sülz, M., Hardebeck, H., Krampitz, G.: Langzeiteinfluß von Cadmiumgaben auf Futterverzehr, Gewichtszunahme, Legeleistung und Eischalenqualität bei Legehennen. Arch. Geflügelkde. *38*, 150 (1974)

386. Hatjipanagiotou, A.: Die Wirkungen von Beryllium-Verbindungen auf das Haushuhn. Dissertation, Bonn 1977

387. Robinson, D. S., King, N. R.: Carbonic anhydrase and formation of the hen's egg shell. Nature *199*, 497 (1963)

388. Robinson, D. S.: The structure of the mammillary layer of the domestic hen's egg shell. Ann. Biol. anim. Biochem. Biophys. *10*, 27 (1970)

389. Grégoire, Ch., Duchâteau, G. H., Florkin, M.: La trame protidique des acre et des perles. Ann. Inst. Océanogr. *31*, 1 (1955)

390. Diamantstein, T., Schlüns, F.: Lokalisation und Bedeutung der Carboanhydrase im Uterus von Legehennen. Acta histochem. *19*, 269 (1964)

391. Diamantstein, T., Bronsch, K., Schlüns, F.: Carbonic anhydrase in the mammillae of the hen's egg shell. Nature *203*, 88 (1964)

392. Diamantstein, T.: Über die lokale Rolle der Carboanhydrase im Hinblick auf die Eischalenverkalkung. Arch. f. Geflügelkde. *20*, 309 (1966)

393. Bernstein, R. S., Nevalainen, J., Schraer, F., Schraer, H.: Intracellular distribution and role of carbonic anhydrase in the avian *(Gallus domesticus)* shell gland mucosa. Biochim. Biophys. Acta *159*, 367 (1968)

394. Schlüns, I., Diamantstein, T.: Behaviour of carbonic anhydrase in the uterus of hens during moult. Nature *209*, 304 (1966)

395. Lutwak-Mann, C.: Carbonic anhydrase in the female reproductive tract. Occurence, distribution and hormonal dependence. J. Endocrin. *13*, 26 (1955)

396. Potz, A.: Personal communication 1978

397. Mongin, P., Etude de l'anhydrase carbonique chez les oiseaux. Annal. Biol. animal. Biochem. Biophys. *10*, 119 (1970)

398. Krampitz, G., Engels, J., Helfgen, I.: Über das Vorkommen von Carboanhydratase in der Eischale des Huhns. Experientia *30*, 228 (1974)

399. Krampitz, G., Engels, J., Heindl, I.: Definition und biologische Bedeutung von Carboanhydratase. Arch. Geflügelkde. *39*, 189 (1975)

400. Engels, J., Heindl. I., Krampitz, G.: Methodisches zur Isolierung von Carboanhydratase aus Eischalen und Erythrocyten des Haushuhns. Arch. Geflügelkde. *40*, 1 (1976)

401. Heindl, I., Engels, J., Krampitz, G.: Eigenschaften von Carboanhydratase aus Eischalen und Erythrocyten des Haushuhns. Arch. Geflügelkde. *40*, 37 (1976)

402. Fridborg, K., Kannan, K. K., Liljas, A., Lundin, J., Strandberg, B., Strandberg, R., Tulander, B., Wiren, G.: Crystal structure of human erythrocyte carbonic anhydrase C. III. Molecular structure of the enzyme and of one enzym-inhibitor complex at 5.5 Å resolution. J. Mol. Biol. *25*, 505 (1967)

403. Riepe, M. E., Wang, J. H.: Infrared studies on the mechanism of action of carbonic anhydrase. J. Biol. Chem. *243*, 2779 (1968)

404. Tyler, C., Simkiss, K.: Some changes in the shell during incubation. J. Sci. Food Agric. *10*, 611 (1959)

405. Schmidt, W. J.: Morphologie der Kalkresorption an der ausgebrüteten Vogeleischale. Z. Zellforsch. *68*, 874 (1965)

406. Terepka, A. R.: Organic-inorganic interrelationships in avian egg shell. Exptl. Cell Res. *30*, 183 (1963)

407. Hamm, M.: Molekularmechanismen der Biomineralisation: Vorkommen und Darstellung Calcium-bindender Systeme. Dissertation, Bonn 1976

408. Hausmanns, G.: Molekularmechanismen der Biomineralisation: Charakterisierung und Biosynthese Calcium-bindender Systeme. Dissertation, Bonn 1977

409. Krampitz, G., Weise, K., Potz, A., Engels, J., Samata, T., Becker, K., Hedding, M., Flajs, G.: Calcium-binding peptide in dinosaur egg shells. Naturwissenschaften *64*, 583 (1977)

410. Krampitz, G., Engels, J., Hamm, M., Kriesten, K., Cazaux, C.: On the molecular mechanism of the biological calcification. 1. Ca-ligands from gastropod shells, egg-shells and uterine fluid of hens. Biomineralisation *9*, 59 (1977)

411. Krampitz, G., Engels, J., Cazaux, C.: Biochemical studies on water-soluble proteins and related components of gastropod shells. In: The mechanism of mineralization in the invertebrates and plants, Vol. 5, p. 155. Watabe, N., Wilbur, K. M. (eds). Columbia, S. C.: The Belle W. Baruch Library in Marine Science University South Carolina Press 1976

412. Krampitz, G., Engels, J., Hamm, M., Samata, T., Cazaux, C.: Biochemical studies on the components of mollusc shells. In: The 3rd Internat. Symp. Mechanisms of Biomineralization in the Invertebrates and Plants, Kashikojima, Japan, 8–11 Oct. 1977; Abstr. No. 16

413. Crenshaw, M. A.: Ionotropic Nucleation by Molluscan Shell Matrix. In: The 3rd Internat. Symp. Mechanisms of Biomineralization in the Invertebrates and Plants, Kashikojima, Japan, 8–11 Oct. 1977; Abstr. No. 32

414. Crenshaw, M. A.: The soluble matrix from *Mercenaria mercenaria* shell. Biomineralisation *6*, 6 (1972)

415. Wilbur, K. M., Simkiss, K.: Calcified shells. In: Comprehensive biochemistry 26A. Florkin, M., Stotz, E. H. (eds.). Amsterdam: Elsevier Publishing Co. 1968

416. Hunt, S.: Polysaccharide-protein complexes in invertebrates. London, New York: Academic Press 1970

417. Tanaka, S., Hatano, H., Suzue, C.: Biochemical studies on pearl. VII. Fractionation and terminal amino acids of conchiolin. J. Biochem. (Tokyo) *47*, 117 (1960)

418. Voss-Foucart, M. F.: Essais de solubilisation et de fractionnement d'une conchioline (nacre murale de *Nautilus pompilius*, mollusque cephalopode). Comp. Biochem. Physiol. *26*, 877 (1968)

419. Meenakshi, V. R., Hare, P. E., Wilbur, K. M.: Amino acids of the organic matrix of neogastropod shells. Comp. Biochem. Physiol. *40B*, 1037 (1971)

420. Voss-Foucart, M. F., Laurent, C., Gregoire, C.: Sur les constituants organiques des coquilles d'etherides. Archs. int. Phys. Biochim. *77*, 901 (1969)

421. Hotta, S.: Infra-red spectra and conformation of protein constituing the nacreous layer of molluscan shell. Earth Sci. Tokyo *23*, 133 (1969)

422. Watabe, N.: Crystal matrix relationship in the inner layers of mollusc shells. J. Ultrastruct. Res. *12*, 351 (1965)

423. Mutvei, H.: Ultrastructure of the mineral and organic components of molluscan nacreous layers. Biomineralisation *2*, 48 (1970)

424. Weiner, S., Hood, L.: Soluble protein of the organic matrix of mollusc shells: A potential template for shell formation. Science *190*, 987 (1975)

425. Weiner, S., Lowenstam, H. A., Hood, L.: Characterization of 80-million-year old mollusc shell proteins. Proc. Nat. Acad. Sci. (Wash.) *73*, 2541 (1976)

426. Samata, T.: personal communication (1978)

427. Sanguansri, P.: Molekularmechanismen der Biomineralisation: Vorkommen, Darstellung und Enzymkinetik von Polyphenoloxidase aus Biomineralisaten. Dissertation, Bonn 1977

428. Degens, E. T., Spencer, D. W., Parker, R. H.: Paleobiochemistry of molluscan shell proteins. Comp. Biochem. Physiol. *201*, 553 (1967)

429. Hunt, S. I., Breuer, S. W.: Chlorinated and brominated tyrosine residues in molluscan sclero-proteins. Biochem. Soc. Trans. *1*, 215 (1973)

430. Crenshaw, M. A., Ristedt, H.: The histochemical localization of reactive groups in septal nacre from *Nautilus pompilius L.* In: The mechanisms of mineralization in the invertebrates and plants. Watabe, N., Wilbur, K. M. (eds). The Belle W. Baruch Library in Marine Science, Vol. 5, pp. 355. University South Carolina Press 1976

431. Pautard, F. G. E.: Calcification in unicellular organisms. In: Biological calcification: Cellular and molecular aspects, pp. 105. Schraer, H. (ed.). Amsterdam: North-Holland Publishing Co. 1970

432. Klaveness, D.: Coccolithus huxleyi (Lohmann) Kamptner: Morfologisk underskelser i lys-og elektronmikroskop. PhD Thesis, Oslo 1971

433. Westbroek, P., De Jong, E. W., Dam, W., Bosch, L.: Soluble intracrystalline polysaccharides from coccoliths of *Coccolithus huxleyi*: (Lohmann) Kamptner I. Calc. Tiss. Res. *12*, 227 (1973)

434. De Jong, L. W., Dam, W., Westbroek, P., Crenshaw, M. A.: Aspects of calcification in *Emiliania huxleyi* (unicellular alga). In: The mechanisms of mineralization in the invertebrates and plants. Watabe, N.,, Wilbur, K. M. (eds.). The Belle W. Baruch Library in Marine Science Vol. 5, pp. 135. University South Carolina Press 1976

435. De Jong, E. W., Bosch, L., Westbroek, P.: Isolation and characterization of a Ca^{2+}-binding polysaccharide associated with coccoliths of *Emiliania huxleyi* (Lohmann) Kamptner. Eur. J. Biochem. *70*, 611 (1976)

436. Woodward, C., Davidson, E. A.: Structure-function relationships of protein polysaccharide complexes: specific ion-binding properties. Proc. Nat. Acad. Sci. (Wash.) *60*, 201 (1968)

437. Towe, K. M.: Invertebrate shell structure and organic matrix concept. Biomineralisation *4*, 1 (1972)

438. Seifert, H.: Strukturgelenkte Grenzflächenvorgänge in der belebten und unbelebten Natur. Naturw. Rundsch. *19*, 1, 50 (1966)

439. Wada, K.: Crystal growth of molluscan shells. Bull. Nat. Pearl Res. Lab, *7*, 703 (1961)

440. Simkiss, K.: The organic matrix of the oyster shell. Comp. Biochem. Physiol. *16*, 427 (1965)

441. Wilbur, K. M., Simkiss, K.: Calcified shells. In: Comprehensive biochemistry, Vol. 26A, pp. 229. Florkin, M., Stotz, E. H. (eds.). Amsterdam: Elsevier 1968

442. Degens, E. T., Matheja, J.: Molecularmechanisms on interactions between oxygen co-ordinated metal polyhedra and biochemical compounds. Techn. Report, Woods Hole Oceanogr. Inst., Ref. No. 57–67 (1967)

443. Erben, H. K.: Über die Bildung und das Wachstum von Perlmutt. Biomineralisation *4*, 15 (1972)

444. Concluding remarks. 3rd Internat. Symp. Mechanisms of Mineralization in the Invertebrates and Plants. Kashikojima, Japan 8–11 Oct. 1977

445. Simkiss, K.: Cellular aspects of calcification. In: The mechanisms of mineralization in the invertebrates and plants. Watabe, N., Wilbur, K. M. (eds.). The Belle W. Baruch Library in Marine Science, Vol. 5, pp. 1, University South Carolina Press 1976

446. Simkiss, K.: Intracellular and extracellular routes in biomineralization. Symp. Soc. Exp. Biol. *30*, 423 (1976)

447. Bryan, G. W.: The occurence and seasonal variation of trace metals in the scallops *Pecten maximus (L)* and *Chlamys opercularis (L)*. J. mar. biol. Ass., U. K. *53*, 145 (1973)

448. Martoja, R., Alibert, J., Ballan-Dufranchais, C., Jeantet, A. J., Lhonore, D., Truchet, M.: Microanalyse et ecologie. J. Microscopic Biol. Cell. *22*, 441 (1975)

449. Von Brand, T. Nylen, M. V. Martin, G. N., Churchwell, K.: Composition and crystallization patterns of calcareous corpuscels of cestodes grown in different classes of hosts. J. Parasit. *53*, 683 (1967)

450. Walker, G., Rainbow, P. S., Foster, P., Crisp, D. J.: Barnacles: possible indicators of zinc pollution? Macr. Biol. *30*, 57 (1975)

451. Jones, A. R.: Mitochondria, calcification and waste disposal. Calc. Tiss. Res. *3*, 363 (1969)

452. Simkiss, K.: Intracellular pH during calcification. A study of the avian shell gland. Biochem. J. *111*, 254 (1969)

453. Humbert, W.: Localization, structure et genèse des concrétions minérales dans le mésenteron des Collemboles Tomoceridae (Insecta, Collembola). Z. Morph. Ökol. Tiere *78*, 93 (1974)

454. Simkiss, K.: Biomineralization and detoxification. Calc. Tiss. Res. *24*, 199 (1977)

455. Anderson, H. C.: Vesicles associated with calcification in the matrix of epiphyseal cartilage. J. Cell. Biol. *41*, 59 (1969)

456. Bonucci, E.: Fine structure and histochemistry of calcifying globules in epiphyseal cartilage. Z. Zellforsch. Mikrosk. Anat. *103*, 192 (1970)

457. Bernard, G. W.: An electron microscopic study of the initial intramembranous ossification. Amer. J. Anat. *125*, 271 (1969)

458. Anderson, C. H.: Calcium-accumulating vesicles in the intercellular matrix of bone. Hard Tissue Growth, Repair and Remineralization. Ciba Foundation Symposium 11 (new series) pp. 213 (1973)

459. Thyberg, J., Friberg, U.: Ultrastructure and phosphatase activity of matrix vesicles and cytoplasmic dense bodies in the epiphyseal plate. J. Ultrastruct. Res. *33*, 554 (1970)

460. Thyberg, J., Friberg, U.: Electron microscopic enzyme. Histochemical studies on the cellular genesis of matrix vesicles in the epiphyseal plate. J. Ultrastruct. Res. *41*, 43 (1972)

461. Anderson, H. C., Matsuzawa, T., Sadjera, S. W., Ali, S. Y.: Membranous particles in calcifying cartilage matrix. Trans. N. Y. Acad. Sci. *32*, 619 (1970)

462. Matsuzawa, T., Anderson, H. C.: Phosphatases of epiphyseal cartilage studied by electron microscopic cytochemical methods. J. Histochem. Cytochem. *19*, 801 (1971)

463. Ali, S. Y., Sajdera, S. W., Anderson, H. C.: Isolation and characterization of calcifying matrix vesicles from epiphyseal cartilage. Proc. Nat. Acad. Sci. (Wash.) *67*, 1513 (1970)

464. Eaton, R. H., Moss, D. W.: Partial purification and some properties of human alkaline phosphatase. Enzymologia *35*, 31 (1968)

465. Perres, N., Sajdera, S. W., Anderson, N. C.: Lipid analysis of vesicles isolated from the matrix of calcifying cartilage. Fed. Proc. *30*, 1244 (1971)

466. Wuthier, R. E., Majeska, R. J., Collins, G. M.: Biosynthesis of matrix vesicles in epiphyseal cartilage. I. In vivo incorporation of ^{32}P orthophosphate into phospholipids of chondrocyte, membrane, and matrix vesicle fractions. Calc. Tiss. Res. *23*, 135 (1977)

467. Wuthier, R. E.: Electrolytes of isolated epiphyseal chondrocates matrix vesicles and extracellular fluid. Calc. Tiss. Res. *23*, 125 (1977)

468. Höhling, H. J., Steffens, H., Stamm, G., Mays, U.: Transmission microscopy of freeze dried, unstained epiphyseal cartilage of the guinea pig. Cell. Tiss. Res. *167*, 243 (1976)

469. Majeska, R. J., Wuthier, R. E.: Studies on matrix vesicles isolated from chick epiphyseal cartilage: Association of pyrophosphatase and ATPase activities with alkaline phosphatase. Biochim. Biophys. Acta *391*, 51 (1975)

470. Felix, R., Fleisch, H.: The Role of Matrix Vesicle in Calcification. Calc. Tiss. Res. *21*, 344 (1976)

471. Lehninger, A. L.: Biochemistry. The Molecular Basis of cell structure and function. New York: Worth Publishers, Inc. 1970

472. Martin, J. H., Matthews, J. L.: Mitochondrial granules in chondrocytes. Calc. Tiss. Res. *3*, 184 (1969)

473. Martin, J. H., Matthews, J. L.: Mitochondrial granules in chondrocytes, osteoblasts and osteocates. Clin. Orthop. *68*, 273 (1970)

474. Matthews, J. L., Martin, J. H., Sampson, H. W., Kunin, A. S., Roan, J. H.: Mitochondrial granules in the normal and rachitic epiphysis. Calc. Tiss. Res. *5*, 91 (1970)

475. Suftin, L. V., Holtrop, M. E., Orgilvie, R. E.: Microanalysis of individual Mitochondrial granules with diameters less than 1.000 Angstroms. Science *174*, 947 (1971)

476. Brighton, C. T., Hunt, R. M.: Mitochondrial calcium and its role in calcification. Histochemical localization of calcium in electron micrographs of the epiphyseal growth plate with K-pyroantimonate. Clin. Orthop. *100*, 406 (1974)

477. Brighton, C. T., Heppenstall, R. B.: Oxygen tension in zones of the epiphyseal plate, the metaphysis and diaphysis. An in vitro and in vivo study in rats and rabbits. J. Bone Joint Surg. *53A*, 719 (1971)

478. Hohman, W., Schraer, H.: The intracellular distribution of calcium in the mucosa of the avian shell gland. J. Cell. Biol. *30*, 317 (1966)

479. Shapiro, I. M., Greenspan, J. S.: Are mitochondria directly involved in biological mineralization? Calc. Tiss. Res. *3*, 100 (1969)

480. Halstead, L. B.: Are mitochondria directly involved in biological mineralization? The mitochondrion and the origin of bone. Calc. Tiss. Res. *3*, 103 (1969)

481. Neuman, W. F., Neuman, M. W.: The nature of mineral phase of bone. Chem. Rev. *53*, 1 (1953)

482. Glimcher, M. J.: Molecular biology of mineralized tissues with particular reference to bone. Rev. modern Phys. *31*, 359 (1959)

483. Höhling, H. J., Ashton, B. A., Köster, H. D.: Quantitative electron microscope investigations of mineral nucleation in collagen. Cell. Tiss. Res. *148*, 11 (1974)

484. Scott. B. L., Pease, D. C.: Electron microscopy of the epiphyseal apparatus. Anat. Rec. *126*, 465 (1956)

485. Robinson, R. A., Cameron, D. A.: An electron microscope study of cartilage and bone matrix at the distal epiphyseal line of the femur in the newborn infant. J. biophys. biochem. Cytol. *2*, 253 (1956)

486. Pautard, F. G. E.: Calcification in single cells: with an appraisal of the relationship between *Spirostomum ambiguum* and the osteocyte. In: The mechanisms of mineralization in the invertebrates and plants. Watabe, N., Wilbur, K. M. (eds.). The Belle W. Baruch Library in Marine Science, Vol. 5, pp. 33, The University of South Carolina Press 1976

487. Ali, S. Y.: Analysis of matrix vesicles and their role in calcification of epiphyseal cartilage. Fed. Proc. *35*, 135 (1976)

488. Bernard, G. W., Pease, D. C.: An electron microscopic study of initial intramembranous osteogenesis. Amer. J. Anat. *125*, 271 (1968)

489. Bonucci, E., Dearden, L. C.: Matrix vesicles in aging cartilage. Fed. Proc. *35*, 163 (1976)

490. Felix, R., Fleisch, H.: Pyrophosphatase and ATPase of isolated cartilage matrix vesicles. Calc. Tiss. Res. *22*, 1 (1976)

491. Ennever, J., Vogel, J. J., Rider, L. J., Boyan-Salyers, B.: Nucleation of microbiological calcification by phospholipid. Proc. Soc. exp. Biol. N. Y. *152*, 148 (1976)

492. Howell, D. S., Pita, J. C., Marguez, J. F., Madruga, J. E.: Partition of calcium phosphate and protein in the fluid phase aspirated at calcifying sites in epiphyseal cartilage. J. Clin. Invest. *47*, 1121 (1968)

493. Höhling, H. J., Steffens, H., Heuck, F.: Untersuchungen zur Mineralisierungsdichte im Hartgewebe mit Protein-Polysaccharid bzw. Kollagen als Hauptbestandteile der Matrix. Z. Zellforsch. *134*, 283 (1972)

494. Höhling, H. J., Neubauer, G., Scholz, F., Boyde, A., Heine, H. G., Reimer, L.: Electronmicroscopical and laser diffraction studies on the nucleation and growth of crystals in the organic matrix of dentine. Z. Zellforsch. *117*, 381 (1971)

495. Höhling, H. J.: Die Bauelemente von Zahnschmelz und Dentin aus morphologischer, chemischer und struktureller Sicht. Habilitationsschrift an der Medizinischen Fakultät der Univ. Münster 1964, München; Hauser, 1966

496. Höhling, H. J., Themann, H., Vahl, J.: Collagen and apatite in hard tissues and pathological formations from a crystal chemical point of view. Calc. Tiss. Proc. 13rd. Europ. Symp. on Calc. Tiss., p. 146. Berlin, Heidelberg, New York: Springer 1966

497. Höhling, H. J., Schöpfer, H.: Morphological investigations of apatite nucleation in hard tissues and salvary stone. Naturwissenschaften 55, 545 (1968)

498. Höhling, H. J.: Collagen mineralization in bone, dentine, cementum and cartilage. Naturwissenschaften 56, 466 (1969)

499. Höhling, H. J., Kreilus, R., Neubauer, G., Boyde, A.: Electron microscopy and electron microscopical measurements of collagen mineralization in hard tissues. Z. Zellforsch. 122, 36 (1971)

500. Smith, W. J.: Molecular patterns in native collagen. Nature 219, 157 (1968)

501. Miller, A., Wray, J. S.: Molecular packing in collagen. Nature 230, 437 (1971)

502. Miller, A., Parry, D. A. D.: The structure and packing of microfibrils in collagen. J. Mol. Biol. 75, 441 (1973)

503. Hulmes, D. J. S., Miller, A., Parry, D. A. D., Piez, K. A., Woodhead-Galloway, J.: Analysis of the primary structure of collagen for the origins of molecular packing. J. Mol. Biol. 79, 137 (1973)

504. Glimcher, M. J.: Specificity of the molecular structure of organic matrices in mineralization. In: Calcification in biologic systems, Wash., pp. 421. Sognnaes, R. F. (ed.). Am. Assoc. Adv. Sci. 1968

505. Jibril, A. O.: Proteolytic degradation of ossifying cartilage matrix and the removal of acid mucopolysaccharides prior to bone formation. Biochim. Biophys. Acta 136, 162 (1967)

506. Bowness, J. M.: Present concepts of the role of ground substance in calcification. Clin. Orthop. 59, 233 (1968)

507. Schubert, M., Pras, M.: Ground substance protein-polysaccharides and the precipitation of calcium phosphate. Clin. Orthop. 60, 235 (1968)

508. Smith, J. W.: Disposition of protein-polysaccharide in the epiphyseal cartilage of the young rabbit. J. Cell. Sci. 6, 843 (1970)

509. Eichelberger, L., Roma, M.: Effects of age on the histochemical characterization of costal cartilage. Amer. J. Physiol. 178, 296 (1954)

510. Howell, D. S., Delchamps, E., Riemer, W., Kiem, K.: A profile of electrolytes in the cartilaginous plate of growing ribs. J. Clin. Invest. 39, 919 (1960)

511. Boyd, E. S., Neuman, W. F.: The ion-binding properties of cartilage. J. Biol. Chem. 193, 243 (1951)

512. Smith, Q. T., Lindenbaum, A.: Composition and calcium-binding of protein-polysaccharides of calf nasal septum and scapula. Calc. Tiss. Res. 7, 290 (1971)

513. Howell, D. S., Pita, J. C.: Calcification of growth plate cartilage with special reference to studies on micropuncture fluids. Clin. Orthop. 118, 208 (1976)

514. Hardingham, T. E., Muir, H.: The specific interaction of hyaluronic acid with cartilage proteoglycans. Biochim. Biophys. Acta 279, 401 (1972)

515. Hascall, V. C., Sajdera, S. W.: Protein-polysaccharide complex from bovine nasal cartilage. The function of glycoprotein in the formation of aggregates. J. Biol. Chem. 244, 2384 (1969)

516. Hascall, V. C. Heinegard, D.: The structure of cartilage proteoglycans. In: Extracellular matrix influences on gene expression, pp. 423. Slavkin, H. C., Greulich, R. C. (eds.). New York: Academic Press 1975

517. Heinegard, D., Hascall, V. C.: Aggregation of cartilage proteoglycans. 3. Characteristics of the proteins isolated from trypsin digests of aggregates. J. Biol. Chem. 249, 4250 (1974)

518. Rosenberg, L., Hellmann, W., Kleinschmidt, A. K.: Electron microscopic studies of proteoglycan aggregates from bovine articular cartilage. J. Biol. Chem. 250, 1877 (1975)

519. Sajdera, S. W., Hascall, V. C.: Protein-polysaccharide-complex from bovine nasal cartilage. A comparison of low and high shear extraction procedures. J. Biol. Chem. 244, 77 (1969)

520. Cuervo, L. A., Pita, J. C., Howell, D. S.: Inhibition of calcium phosphate mineral growth by proteoglycan aggregate fractions in a synthetic lymph. Calc. Tiss. Res. *13*, 1 (1973)

521. Pita, J. C., Howell, D. S., Kuettner, K.: Evidence for a role of lysozyme in endochondral calcification during healing of ricketts. In: Extracellular matrix influence on gene expression. Slavkin, H. C., Greulich, R. E. (eds.). New York: Academic Press 1975

522. Fleisch, H., Felix, R., Hansen, T., Schenk, R.: Role of the organic matrix in calcification. In: Extracellular matrix influences on gene expression. Slavkin, H. C., Greulich, R. E. (eds.). New York: Academic Press 1975

523. Engfeldt, B., Hulth, A., Westerborn, O.: Effect of papain on bone. I. A histologic, autoradiographic, and microradiographic study on young dogs. A. M. A. Arch. Pathol. *68*, 600 (1959)

524. Ali, S. Y., Evans, L.: Studies on the cathepsins in elastic cartilage. Biochem. J. *112*, 427 (1969) (1969)

525. Barrett, A. J.: Cathepsin D. Purification of isoenzymes from human and chicken liver. Biochem. J. *117*, 601 (1970)

526. Fell, H. B., Dingle, T. J.: Studies on the made of action of excess vitamin A. 6. Lysosomal protease and the degradation of cartilage matrix. Biochem. J. *87*, 403 (1963)

527. Sapolsky, A. I., Altman, R. D., Woessner, J. F., Howell, D. S.: The action of cathepsin D in human articular cartilage on proteoglycans, J. Clin. Invest. *52*, 624 (1973)

528. Woessner, J. F. Jr., Shamberger, R. J. Jr.: Purification and properties of cathepsin D from bovine uterus. J. Biol. Chem. *246*, 1951 (1971)

529. Sapolsky, A. I., Howell, D. S., Woessner, J. F. Jr.: Neutral proteases and cathepsin D in human articular cartilage. J. Clin. Invest. *53*, 1044 (1974)

530. Sapolsky, A. I., Howell, D. S., Woessner, J. F. Jr.: Proteoglycan degradation by neutral protease and acid cathepsin in human articular cartilage: inhibition by *D*-penicillamine. Arthr. Rheum. *18*, 423 (Abstr.) (1975)

531. McCord, J. M.: Free radicals acid inflammation: Protection of synovial fluid by superoxide dismutase. Science *185*, 529 (1974)

532. Wadkins, C. L., Luben, R., Thomas, M., Humphreys, R.: Physical biochemistry of calcification. Clin. Orthop. *99*, 246 (1974)

533. Fleisch, H., Russell, R. G. G.: A review of the physiological and pharmacological effects of pyrophosphate and diphosphonates on bone and teeth. J. Dent. Res. *51*, 324 (1972)

534. Russell, R. G. G., Smith, G.: Diphosphonates: experimental and clinical aspects. J. Bone Jt. Surg. *55B*, 66 (1973)

535. Russell, R. G. G. Fleisch, H.: Pyrophosphate and diphosphonates in skeletal metabolism. Physiological, chemical, and therapeutical aspects. Clin. Orthop. *108*, 241 (1975)

536. Russell, R. G. G., Mühlbauer, R. C., Bisaz, S., Williams, D. A., Fleisch, H.: The influence of pyrophosphate, condensed phosphates, phosphonates and other phosphate compounds on the dissolution of hydroxyapatite in vitro and on bone resorption induced by parathyroid hormone in tissue culture and in thyroparathyroidectomized rats. Calc. Tiss. Res. *6*, 183 (1970)

537. Jung, A. Bisaz, S., Fleisch, H.: The binding of pyrophosphate and two diphosphonates by hydroxyapatite crystals. Calc. Tiss. Res. *11*, 269 (1973)

538. Fleisch, H., Russell, R. G. G., Bisaz, S., Casey, P. A., Mühlbauer, R. C.: The influence of pyrophosphate analogues (diphosphonates) on the precipitation and dissolution of calcium phosphate in vitro. Calc. Tiss. Res. *2*, 10–10A (1968)

539. Fleisch, H., Russell, R. G. G., Francis, M. D.: Diphosphonates inhibit hydroxyapatite dissolution in vitro and bone resorption in tissue culture and in vivo. Science *165*, 1262 (1969)

540. Reynolds, J. J., Minkin, C., Morgan, D. B., Spycher, D., Fleisch, H.: The effects of two diphosphonates on the resorption of mouse calvaria in vitro. Calc. Tiss. Res. *10*, 302 (1972)

541a. Michael, W. R., King, W. R., Francis, M. D.: Effectiveness of dophosphonates in preventing "osteoporosis" of disuse in the rat. Clin. Orthop. *78*, 271, (1971)

541b. Gasser, A. B., et al. The influence of two diphosphonates on calcium metabolism in the rat. Clin. Sci. *43*, 31 (1972)

542. Schenk, R., et al.: Effect of ethane-1-hydroxy-1, 1-diphosphonate (EHDP) and dichlormethylene diphosphonate (Cl$_2$ MDP) on the calcification and resorption of cartilage and bone in the tibial epiphysis and metaphysis of rats. Calc. Tiss. Res. *11*, 196 (1973)

543. King, W. R., Francis, M. D., Michael, W. R.: Effect of disodium-ethane-1-hydroxy-1, 1-diphosphonate on bone formation. Clin. Orthop. *78*, 251 (1971)

544. Fleisch, H., Russell, R. G. G., Bisaz, S., Mühlbauer, R. C., Williams, D. A.: The inhibitory effect of phosphonates on the formation of calcium phosphate crystals in vitro and on aortic and kidney calcification in vivo. Europ. J. Clin. Invest. *1*, 12 (1970)

545. Miller, S. C., Jee, W. S. S.: Ethane-1-hydroxy-1, 1-diphosphonate (EHDP) effects on growth and modeling of the rat tibia. Calc. Tiss. Res. *18*, 215 (1975)

546. Minkin, C., Rabadjija, L., Goldhaber, P.: Bone remodeling in vitro: The effects of two diphosphonates on osteoid synthesis and bone resorption in mouse calvaria. Calc. Tiss. Res. *14*, 161 (1974)

547. Goulding, A., Chesner, R.: Comparison of effects of parathyroidectomy and ethane-1-hydroxy-1, 1-diphosphonate (EHDP) administration upon bone synthesis of hydroxyproline and urinary excretion of hydroxyproline in rats. Calc. Tiss. Res. *23*, 115 (1977)

548. Miller, S. C., Jee, W. S. S.: The comparative effects of dichloromethylene diphosphonate (Cl_2 MDP) and ethane-1-hydroxy-1, 1-diphosphonate (EHDP) on growth and modeling of the rat tibia. Calc. Tiss. Res. *23*, 207 (1977)

549. Münzenberg, K. J., Reimann, E., Gebhardt, M., Kühne, H.: Influence of methan-diphosphonate on crystallization in vitro. Biomineralisation *8*, 9 (1975)

550. Fast, D. K., Felix, R., Neuman, W. F., Sallis, J., Fleisch, H.: The effects of diphosphonates on cells in culture. Calc. Tiss. Res. *22*; 449 (1977)

551. Bonjour, J. P., Russell, R. G. G., Morgan, D. B., Fleisch, H. A.: Intestinal calcium absorption, Ca-binding protein, and Ca-ATPase in diphosphonate-treated rats. Amer. J. Physiol. *224*, 1011 (1973)

552. Morgan, D. B., Bonjour, J. P., Gasser, A. B., O'Brien, K., Fleisch, H.: The influence of a diphosphonate on the intestinal absorption of calcium. Israel J. med. Sci. *7*, 384 (1971)

553. Baxter, L. A., De Luca, H. F., Bonjour, J. P., Fleisch, H.: Inhibition of vitamin D metabolism by ethane-1-hydroxy-1, 1-diphosphonate. Arch. Biochem. *164*, 655 (1974)

554. Hill, L. F., Lumb, G. A., Maver, G. A., Stanbury, S. W.: Indirect inhibition of the biosynthesis of 1,25-dihydroxycholecalciferol in rats treated with a diphosphonate. Clin. Sci. *44*, 335 (1973)

555. Taylor, C. M., Mawer, E. B., Reeve, A.: The effects of diphosphonate and dietary calcium on the metabolism of vitamin D_3 (cholecalciferol) in the chick. Clin. Sci. Molec. Med. *49*, 391 (1975)

556. Bonjour, J. P., De Luca, H. F., Fleisch, H., Trechsel, U., Matejowec, L. A., Omdahl, J. L.: Reversal of the EHDP inhibition of calcium absorption by 1,25-dihydroxycholecalciferol. Europ. J. Clin. Invest. *3*, 44 (1973)

557. Bonjour, J. P., Trechsel, U., Fleisch, H., Schenk, R., De Luca, H. F., Baxter, L. A.: Action of 1,25-dihydroxyvitamin D_3 and a diphosphonate on calcium metabolism in rats. Amer. J. Physiol. *229*, 402 (1975

558. Henry, H. L., Norman, A. W.: Biochemical and physiological regulation of 25-hydroxycholecalciferol-1-hydroxylase. In: Vitamin D and problems related to uremic bone disease, pp. 6. Norman, A. W., Schaefer, K., Grigoleit, H. G., Herrath, D. V., Ritz, E. (eds.). Berlin: de Gruyter 1975

559. Trechsel, U., Bonjour, J. P., Fleisch, H.: Influence of dietary calcium, phosphorus, and vitamin D on the conversion of 25-hydroxyvitamin D_3 to 1,25-dihydroxy-vitamin D_3 by kidney tubules of diphosphonate-treated quails. Calc. Tiss. Res. *22*, 461 (1977)

560. Watabe, N.: Crystallographic analysis of the coccolith *Coccolithus huxleyi*. Calc. Tiss. Res. *1*, 114 (1967)

561. Pautard, F. G. E.: Calcification in unicellular organisms. In: Biological calcification. Cellular and molecular aspects, pp. 105. Schraer, H. (ed.). New York: Appleton-Century Crofts 1970

562. Mantun, I.: Further observations on scale formation in *Clerysochromulina chiton*. J. Cell. Sci. *2*, 411 (1967)

563. Wilbur, K. M., Watabe, N.: Experimental studies on calcification in molluscs and the alga *Coccolithus huxleyi*. Ann. N. Y. Acad. Sci. *109*, 82 (1963)

G. Krampitz and W. Witt

564. Manton, I., Leedale, G. F.: Observations on the fine structure of *Paraphysomonas vestita*, with special reference to the Golgi apparatus and the origin of scales. Phycologia *1*, 37 (1961)

565. Brown, R. M., Herth, W., Franke, W. W., Romanovicz, D.: The role of the Golgi apparatus in the biosynthesis and secretion of a cellulosic glycoprotein in Pleurochrysis: a model system for the synthesis of structural polysaccharides. In: Biogenesis of plant cell wall polysacchari-des, pp. 207. Loewus, F. (ed.). New York: Academic Press 1973

566. Pautard, F. G. E.: Hydroxyapatite as a development feature of *Spirostonum ambiguum*, Biochim. Biophys. Acta *35*, 35 (1959)

567. Osvoji, C. I., Rowless, S. L.: Studies on the organic composition of dental calculus and related calculi. Calc. Tiss. Res. *16*, 193 (1974)

568. Lustmann, J., Lewin-Epstein, J., Shteyer, A.: Scanning electron microscopy of dental cal-culus. Calc. Tiss. Res. *21*, 47 (1976)

569. Francis, M. D., Briner, W. W.: The effect of phosphonates on dental enamel in vitro and calculus formation in vivo. Calc. Tiss. Res. *11*, 1 (1973)

570. Briner, W. W., Francis, M. D.: In vitro and in vivo evaluation of anti-calculus agents. Calc. Tiss. Res. *11*, 10 (1973)

571. Goreau, T. F.: Problems of growth and calcium deposition in reef corals. Endeavour *20*, 32 (1961)

572. Goreau, T. F.: Histochemistry of mucopolysaccharide-like substances and alkaline phos-phatase in Madreporaria. Nature *177*, 1029 (1956)

573. Goreau, T. F.: The physiology of skeleton formation in corals. 1. A method for measuring the rate of calcium deposition by corals under different conditions. Biol. Bull. mar. biol. Lab. Woods Hole *116*, 59 (1959)

574. Wilbur, K. M.: Shell formation and regeneration. In: Physiology of mollusca, pp. 243. Wilbur, K. M., Yonge, C. M. (eds.). New York: Academic Press 1964

575. Wilbur, K. M.: Shell formation in molluscs. Chemical Zoology *7*, 103 (1972)

576. Travis, D. F.: The comparative ultrastructure and organization of five calcified tissues. In: Biological calcification: Cellular and molecular aspects, pp. 203. Schraer, H. (ed.). Amster-dam: North Holland Publishing Co. 1970

577. Neuman, W. F., Neuman, M. W.: The chemical dynamics of bone mineral. Chicago: Uni-versity Press 1958

578. Kobayashi, S.: Acid mucopolysaccharides in calcified tissues. Int. Rev. Cytol. *30*, 257 (1971)

579. Pautard, F. G. E.: A biomolecular survey of calcification. In: Calcified tissues, pp. 108. Fleisch, H., Blackwood, H. J. J., Owen, M. (eds.). New York, Heidelberg, Berlin: Springer 1966

580. Simkiss, K., Tyler, C.: Reactions between egg shell matrix and metallic ions. Q. J. Microsc. Sci. *99*, 5 (1958)

581. Simkiss, K.: Some properties of the organic matrix of the shell of the cockle. Proc. malac. Soc. Lond. *34*, 89 (1960)

582. Meenakshi, V. R., Donnay, G., Blackwelder, P. L., Wilbur, K. M.: The influence of substrate on calcification patterns in molluscan shell. Calc. Tiss. Res. *15*, 31 (1974)

583. Carter, M. J.: Carbonic anhydrase: isoenzymes properties distribution and functional signifi-cance. Biol. Rev. *47*, 465 (1972)

584. Saleuddin, A. S. M.: Isoenzymes of alkaline phosphatase in *Anodonta grandis* during shell regeneration. Malacologia *9*, 501 (1969)

585. Bradfield, J. R. G.: The localization of enzymes in cells. Biol. Rev. *25*, 113 (1950)

586. Kroon, D. B.: Phosphatase and the formation of protein-carbohydrate complexes. Acta anal. *15*, 317 (1952)

587. Simkiss, K.: Phosphates as crystal poisons. Biol. Rev. *39*, 487 (1964)

588. Guardabassi, A.: Le caratteristicke citologiche ultrastructurali, iscochime e biochemiche di alcumi tessuti e organi che partecipano alla alaborazione di structure calcificate. Monitore zool. ital. *70*, 1 (1962)

589. Schatzmann, H. J.: Dependence on calcium concentration and stoichiometry of the calcium pump in human red cells. J. Physiol. (Lond.) *235*, 551 (1973)

590. Irving, J. T.: Calcium and Phosphorus Metabolism. New York: Academic Press 1973

591. Baker, P. F.: Transport and metabolism of calcium ions in nerve. Prog. Biophys. molec. Biol. *24*, 177 (1972)

592. Crenshaw, M. A.: The inorganic composition of molluscan extrapallial fluid. Biol. Bull. mar. bioL Lab., Woods Hole *143*, 506 (1972)
593. Potts, W. T. W.: The inorganic composition of the blood of *Mytilus edulis* and *Anodonta cygnea.* J. exp. Biol. *31*, 376 (1954)
594. Burton, R. F.: The storage of calcium and magnesium phosphate and of calcite in the digestive glands of the Pulmonata. Comp. Biochem. Physiol. *43*, 655 (1972)
595. Istin, M. Kirschner, L. B.: On the origin of the bioelectrical potential generated by the freshwater clam mantle. J. gen. Physiol. *51*, 478 (1968)
596. Istin, M., Fossat, B.: Etude du profile de potential electrique de la partie centrale du manteau de la moule d'eau douce. C. r. hebd. Sèanc. Acad. Sci., Paris *274*, 119 (1972)
597. Istin, M., Maetz, J.: Perméabilité au calcium manteau de lamellibranches d'eau douce étudiée à l'aide des isotopes ^{45}Ca and ^{47}Ca. Biochim. Biophys. Acta *88*, 225 (1964)
598. Greenaway, P.: Calcium regulation in the freshwater mollusc *Limnea stagnalis.* 1. The effect of internal and external calcium concentration. J. exp. Biol. *54*, 199 (1971)
599. Robertson, J. D.: Ionic regulation in the crab *Carcinus maenas* in relation to the moulting cycle. Comp. Biochem. Physiol. *1*, 183 (1960)
600. Hurwitz, S., Cohen, I., Bar, A.: The transmembrane electrical potential difference in the uterus (shell gland) of birds. Comp. Biochem. Physiol. *35*, 873 (1970)
601. Simkiss, K.: Intracellular pH during calcification. Biochem. J. *111*, 647 (1969)
602. Lörcher, K., Hodges, R. D.: Some possible mechanism of formation of the carbonate fraction of egg shell calcium carbonate. Comp. Biochem. Physiol. *28*, 119 (1969)
603. Campbell, J. W., Speeg, K. V.: Ammonia and biological deposition of calcium carbonate. Nature *224*, 725 (1969)
604. Reddy, G., Campbell, J. W.: Correlation of NH_3 liberation and Ca-deposition by the avian egg and blood NH_3 levels in the laying hen. Experientia *28*, 530 (1972)
605. Digby, P. S. B.: Mechanisms of calcification in mammalian bone. Nature *212*, 1250 (1966)
606. Digby, P. S. B.: Calcification and its mechanisms in the shore crab *Carcinus maenas.* Proc. Linn. Soc. Lond. *178*, 129 (1967)
607. Digby, P. S. B.: Mobility and crystalline form of the lime in the cuticle of the shore crab *Carcinus maenas.* J. ZooL Lond. *154*, 273 (1968)
608. Digby, P. S. B.: The mechanism of calcification in the molluscan shell. Symp. Zool. Soc. Lond. *22*, 93 (1968)
609. Greenaway, P.: Calcium balance at the past moult stage of the freshwater crayfish *Austro-potamoblus pallipes.* J. exp. Biol. *61*, 35 (1974)
610. Jahn, T. L.: A possible mechanism of the effect of electrical potentials on apatite formation in bone. Clin. Orthop. *56*, 261 (1968)
611. Diamond, J. H., Tormey, J. H.: Role of long extracellular-channels in fluid transport across epithelia. Nature *210*, 817 (1966)
612. Diamond, J. M., Bossert, W. H.: Stamding gradient osmotic flow. A mechanism of coupling of water and solute transport in epithelia. J. gen. Physiol. *50*, 2061 (1967)
613. Van Gansen, P. S.: Structure de glandes calciques d'*Eisenia foetida Sav.* Bull. biol. Fr. Belg. *93*, 38 (1959)
614. Nakahara, H., Bevelander, G.: An electron microscope and autoradiographic study of the calciferous glands of the earthworm *Lumbricus terrestris.* Calc. Tiss. Res. *4*, 193 (1969)
615. Leonard, F., Scullin, R. I.: New mechanisms for calcification of skeletal tissues. Nature *224*, 1113 (1969)
616. Russell, R. G. G., Monod, A.: Bonjour, J. P., Fleisch, H.: Relation between alkaline phosphatase and Ca^{2+} ATPase in calcium transport. Nature. New Biol. *240*, 126 (1972)
617. Hasselbach, W., Makinose, M.: The reserval of the sarcoplasmic calcium pump. In: Role of membranes in secretory processes, pp. 158. Bolus, L., Keynes, R. D., Wilbrandt, W. (eds.). Proc. Intern. Conf. Biol. Membranes. Amsterdam: North Holland 1972
618. Lehninger, A. L.: Metabolite carriers in mitochondrial membranes. In: The Ca^{++} transport system in the dynamic structure of cell membranes, pp. 119. Wallach, D. F. H., Fisher, H. (eds.). New York, Heidelberg, Berlin: Springer 1971
619. Martin, J. H., Matthews, J. L.: Mitochondrial granules in chondrocytes. Calc. Tiss. Res. *3*, 184 (1969)

G. Krampitz and W. Witt

620. Lehninger, A. L.: Mitochondria and calcium transport. Biochem. J. *119*, 129 (1970)

621. Shapiro, I. M., Greenspan, J. S.: Are mitochondria directly involved in biological mineralization? Calc. Tiss. Res. *3*, 100 (1969)

622. Elder, J. A., Lehninger, A. L.: Respiration dependent transport of carbon dioxide into rat liver mitochondria. Biochemistry *12*, 976 (1973)

623. Crang, R. E., Holsen, R. C., Hitt, J. B.: Calcite production in mitochondria of earthworm calciferous glands. Bioscience *18*, 299 (1968)

624. Gouranton, J.: Composition, structure et mode de formation des concretions minerales dans l'intestin moyen des homopteres cereopides. J. Cell. Biol. *37*, 316 (1968)

625. Nieland, M. L., Von Brand, T.: Electron microscopy of cestode calcareous corpuscle formation. Expl. Parasit *24*, 279 (1969)

626. Wilbur, K. M., Watabe, N.: Experimental studies on calcification in molluscs and the alga *Coccolithus huxleyi*. Ann. N. Y. Acad. Sci. *109*, 82 (1963)

627. Borle, A. B.: Calcium metabolism at the cellular level. Fed. Proc. Fedn. Amer. Socs. exp. Biol. *32*, 1944 (1973)

628. Kawaguti, S., Kamishima, Y.: Electron microscopy on a gorgonian coral *Anthoplexaura dimopha*. Publs. Seto mar. biol. Lab. *20*, 785 (1973)

629. Crenshaw, M. A., Heeley, J. D.: Sudanophilia at sites of calcification in molluscs. 45th Gen. Meet. Int. Am. Dent. Res. Abst. *119*, 65 (1967)

630. Travis, D. F.: Matrix and mineral deposition in skeletal structures of the decapod crustacea. In: Calcification in biological systems, pp. 57. Sognnaes, R. F. (ed.). Washington: A. A. A. S. 1960

631. Schraer, R., Elder, J. A., Schraer, H.: Aspects of mitochondrial function in calcium movement and calcification. Fed. Proc. Fedn. Amer. Soc. exp. Biol. *32*, 1938 (1973)

632. Simkiss, K.: Cellular aspects of calcification. In: The mechanisms of mineralization in the invertebrates and plants, pp. 1. Watabe, N., Wilbur, K. M. (eds.). The Belle W. Baruch Library in Marine Science, Vol. 5. The University South Carolina Press 1976

633. Von Brandt, T., Nylen, M. U., Martin, G. N., Churchwell, F. K.: Composition and crystallization patterns of calcerous corpuscles of cestodes grown in different classes of hosts. J. Parasit. *53*, 683 (1967)

634. Graf, F.: Dynamique de calcium dans l'epithelium des Caecum posterieurs d'*Orchestria corimana*. C. r. hebd. Seanc. Acad. Sci., Paris *273*, 1828 (1971)

635. Burton, R. F.: Calcium balance and acid-base balance in *Helix aspersa*. In: Perspectives in experimental biology, pp. 7. Davies, S. (ed.). Oxford: Pergamon Press 1976

636. Götting, K. J.: Malakozoologie: Grundriß der Weichtierkunde, pp. 1/39. Stuttgart: Verlag Gustav Fischer 1974

637. Hauschka, P. V., Gallup, P. M.: Purification and calcium-binding-properties of osteocalcin, the γ-carboxyglutamate-containing-protein of bone. In: Calcium binding proteins and calcium function, pp. 338. Wassermann, R. H. et al. New York: North Holland 1977

638. Price, P. A., Otsuka, A. S., Poser, J. W.: Comparison of γ-carboxyglutamic acid-containing proteins from bovine and swordfish bone: Primary structure and Ca^{++} binding, In: calcium binding protein and calcium function, p.333. Wassermann, R. A., Corradino, E., Carafoli, R. H., Kretsinger, D. H., Lemman, Mac, Siegel, F. L. (eds.). New York: North Holland 1977

639. King, Jr., K.: γ-Carboxyglutamic acid in fossil bones and its significance for amino acid dating. Nature *273*, 41 (1978)

640. Dimuzio, M. T., Veis, A.: Phosphoryns-major noncollagenous proteins of rat incisor dentin. Calc. Tiss. Res. *25*, 169 (1978)

641. Höhling, H. J., Steffens, H., Ashton, B. A., Nicholson, A. A. P.: 2. Molekularbiologie der Hartgewebsbildung (Referat). Verh. Dtsch. Ges. Path. *58*, 54 (1974)

642. Thyberg, J.: Electron microscopic studies on the initial phases of calcification in Guinea pig epiphyseal cartilage. J. Ultrastruct. Res. *46*, 206 (1974)

643. Miller, A., Parry, D. A. D.: The structure and packing of microfibrils in collagen. J. Molec. Biol. *75*, 441 (1973)

Received October 23, 1978

Oxygenases and Dioxygenases

Mitsuhiro Nozaki

Department of Biochemistry, Shiga University of Medical Science, Seta, Otsu, Shiga 520-21, Japan

Table of Contents

M. Nozaki

I. Introduction

Respiration is one of the most fundamental activities of life. In mammals, respiration can be divided into two parts, external respiration and internal respiration. Air is introduced into the lung by external respiration and oxygen in the air is then tranferred to various tissues by the action of hemoglobin and myoglobin. Utilization of the oxygen in these tissues is referred to as internal respiration. Then, what is the role of oxygen in internal respiration? Molecular oxygen in tissues serves two functions; one, to act as ultimate hydrogen acceptor in the process of the biological oxidation of foodstuff being reduced to either water or hydrogen peroxide, and two, to transform dietary nutrients into cellular constituents and biologically important substances. The enzymes involved in the former process are usually referred to as oxidases, whereas those involved in the latter process are referred to as oxygenases[1].

In 1955, two types of oxygenases were independently discovered by two groups of investigators. Mason and his collaborators found that during the oxidation of 3,4-dimethylphenol to dimethylcatechol by phenolase, the oxygen atom incorporated into the substrate was derived exclusively from molecular oxygen[2] [Eq. (1)].

$$\tag{1}$$

On the other hand, Hayaishi and his associates found that the two atoms of oxygen incorporated into catechol by the action of pyrocatechase were both derived from molecular oxygen[3] [Eq. (2)].

$$\tag{2}$$

In the case of an oxidase reaction, oxygen serves as an hydrogen acceptor and is reduced to either H_2O [Eq. (3)] or H_2O_2 [Eq. (4)], where SH_2 represents substrate.

$$SH_2 + 1/2\, O_2 \longrightarrow S + H_2O \tag{3}$$

$$SH_2 + O_2 \longrightarrow S + H_2O_2 \tag{4}$$

In contrast, in the above reactions given in Eqs. (1) and (2), either one or two atoms of molecular oxygen are incorporated into their substrates. Therefore, the enzymes that catalyze such oxygen-fixation reactions are termed "oxygenases"[4].

Since the discovery of oxygenase, many oxygenases have been found to be widely distributed in nature and to play physiologically important roles including the biosynthesis, transformation and degradation of phenol compounds, amino acids, lipids, vitamins, etc., as well as the metabolic disposal of a variety of drugs and foreign compounds. For the physiological significance and details of oxygenases, the readers are referred to monographs[5-7] or other review articles[8-14].

II. Classification of Oxygenases

As in the cases of phenolase and pyrocatechase, respectively, oxygenases catalyze the incorporation of either one or two atoms of molecular oxygen into their substrates. Therefore, they are classified into two major groups, monooxygenases and dioxygenases[15].

A) Monooxygenases

Monooxygenases are defined as a group of enzymes that catalyze the incorporation of one atom of molecular oxygen into a substrate. In monooxygenase reactions, molecular oxygen (O_2) reacts with the substrate and one atom of oxygen is incorporated into the substrate, whereas the other atom of oxygen is reduced to water by a reductant, which is essential for the reaction as shown in Eq. (5), where H_2X and S denote the reducing agent and substrate, respectively.

$$S + O_2 + H_2X \longrightarrow SO + H_2O + X \tag{5}$$

Since both oxygenation and oxidation reactions are involved in the reaction, monooxygenases are sometimes referred to as "mixed function oxidases" [8]. The term "hydroxylase" is also used for some monooxygenases which catalyze the formation of a hydroxyl group as a result of monooxygenation. However, not all hydroxylation reactions are catalyzed by monooxygenases. Dioxygenases sometimes catalyze the hydroxylation reaction as will be discussed later and some hydroxyl groups may not be derived from molecular oxygen but from water. For this reason, the term "hydroxylase" is rather misleading.

Hydrogen (or electron) donors, coenzymes and electron carriers involved in the monooxygenase reactions are listed in Table 1. In some monooxygenase reactions, the substrate itself serves as an electron donor as well as an oxygen acceptor [Eq. (6)],

$$SH_2 + O_2 \longrightarrow SO + H_2O \tag{6}$$

where SH_2 denotes the substrate. These types of enzymes may be referred to as *"internal monooxygenases"*. Most of the other monooxygenases, however, require various kinds of external hydrogen (or electron) donors; therefore, they are referred to as *"external monooxygenases"*.

All the internal monooxygenases that have so far been purified and characterized contain flavin coenzymes. The external hydrogen donors include reduced NAD, reduced NADP, ascorbic acid and sulfhydryl compounds. Cofactors required for the external monooxygenases are flavin, pteridine, copper, nonheme iron and heme as cytochrome P-450. In some monooxygenase reactions, enzymes and/or electron carrier systems other than monooxygenase itself are involved in the transfer of an electron or hydrogen from the external hydrogen donor to the cofactor involved.

On the other hand, in the fatty acid desaturation reaction, molecular oxygen is not incorporated into the final product. Nevertheless, monooxygenation is believed

Table 1. Electron donors, electron transport systems and cofactors involved in monooxygenase reaction[a]

Electron donors	Electron transport systems[b]	Cofactors	Examples	Ref.
Substrate	—	FAD (FMN)	Lysine monooxygenase	16–18)
NAD(P)H	—	FAD	Salicylate monooxygenase	19, 20)
Ascorbic acid	—	Cu	Dopamine β-hydroxylase	21–23)
NADPH	Dihydrobiopterine reductase	Biopterine-NHI	Phenylalanine hydroxylase	24–27)
NADPH	NADPH-cyt. P450 reductase (fp)	P-450 (Hp)	Liver microsomal drug hydroxylase	28–32)
NADH	NADH-cyt b_5 reductase (fp), cyt. b_5 (Hp)	NHI	Liver microsomal fatty acid desaturase	33–35)
NADPH	NADPH-adrenodoxin reductase (fp), adrenodoxin (Fe-S)	P-450	Adrenal mitochondrial steroid hydroxylase	36–39)
NADH	NADH-putidaredoxin reductase (fp), putidaredoxin (Fe-S)	P-450	Pseudomonas putida D-camphor hydroxylase	40–42)
NADH	NADH-rubredoxine reductase (fp), rubredoxin (NHI)	NHI	Pseudomonas oleovorans fatty acid-ω-hydroxylase	43–45)
NADH	NHI-FAD	NHI	p-Methylbenzoate demethylase	46, 47)
NADPH	?	FAD	Liver microsomal amino oxidase	48)
NADPH	?	NHI-FAD	Squalene epoxidase	49)
NAD(P)H	?	P-420 (Hp)	Pseudomonas aminovolans secondary amino oxidase	50)

[a] The following abbreviations are used;
fp: flavoprotein; NHI: nonheme iron protein, Hp: hemoprotein,
Fe-S: iron-sulfur protein.

[b] A bar represents that electron carriers other than monooxygenases are not involved in the reaction and electron donors directly reduce cofactors involved in the monooxygenase. A question mark represents that an involvement of electron transport system is not known yet.

M. Nozaki

to be involved in the reaction, since the reaction requires oxygen and NADPH. As shown in Eq. (7), hydroxy fatty acid is proposed to be an enzyme-bound intermediate of the reaction[51, 52].

$$R-CH_2-CH_2-R' \xrightarrow{\quad O_2 + NADPH + H^+ \quad NADP^+ + H_2O \quad} \left[R-CH_2-\underset{OH}{CH}-R' \right] \tag{7}$$

$$\xrightarrow{\quad H_2O \quad} R-CH=CH-R'$$

B) Dioxygenases

Dioxygenases are defined as a group of enzymes that catalyze the incorporation of two atoms of molecular oxygen into the substrate. In most cases, one substrate acts as an oxygen acceptor and a single molecule of the substrate receives 2 atoms of oxygen as shown in Eq. (8),

$$S + O_2 \longrightarrow SO_2 \tag{8}$$

where S donates substrate.

In some dioxygenase reactions, however, one atom each of oxygen molecule is incorporated into two different molecules of one substrate [Eq. (9)] or [Eq. (10)] into two different substrate molecules

$$2 S + O_2 \longrightarrow 2 SO \tag{9}$$
$$S + S' + O_2 \longrightarrow SO + S'O \tag{10}$$

The term "intramolecular dioxygenases" may be used for the dioxygenases catalyzing the reaction shown in Eq. (8), and "intermolecular dioxygenases" for those catalyzing the reactions shown in Eqs. (9) and (10), respectively.

In the reaction represented in Eq. (10), one of the two substrates is invariably α-ketoglutarate, which is converted to succinate by the incorporation of one atom of oxygen with concomitant decarboxylation. Therefore, the overall reaction may be schematically shown as in Eq. (11).

$$\begin{array}{l} COOH \\ | \\ CH_2 \\ | \\ CH_2 \\ | \\ C=O \\ | \\ COOH \end{array} + O_2 + S \longrightarrow \begin{array}{l} COOH \\ | \\ CH_2 \\ | \\ CH_2 \\ | \\ COOH \end{array} + CO_2 + SO \tag{11}$$

α-Ketoglutarate   Succinate

150

Table 2. Dioxygenases and their cofactors

Reactions in which dioxygenases are involved	Cofactors	Examples	Ref.
Ring-cleavage reaction	Fe(II)	Metapyrocatechase	53, 54)
	Fe(III)	Pyrocatechase	55, 56)
	Heme	Tryptophan 2,3-dioxygenase	57, 58)
	Cu	Quercetinase	59, 60)
	FAD	2-Methyl-3-hydroxy-pyridine 5-carboxylate dioxygenase	61)
Double-hydroxylation reaction	Non-heme iron NAD(P)H, FAD	Benzoate hydroxylase	62)
α-Ketoglutarate-requiring reaction	Fe(II), ascorbic acid	γ-Butyrobetaine dioxygenase	63–65)
Oxygenation reaction of sulfur and sulfur compound	GSH	Sulfur oxygenase	66, 67)
	Fe(III), sulfide	Cysteamine oxygenase	68–70)
	NAD(P)H	Cystein oxygenase	71–73)
	Fe(II), unidentified factor		
Miscellaneous	Fe(II)	Lipoxygenase	74–76)
	Heme, tryptophan	Prostaglandin synthetase	77–80)
	GSH		

The dioxygenases involved in different type of reactions and their cofactors are summarized in Table 2. A major function of dioxygenases is the cleavage of the aromatic ring with the insertion of two atoms of molecular oxygen. The indole ring-cleaving enzyme, tryptophan 2,3-dioxygenase contains heme as a cofactor[58]. A flavonol-cleaving enzyme, quercetinase, has been reported to be a copper protein[60] and a pyridine ring-cleaving enzyme, 2-methyl-3-hydroxypyridine-5-carboxylate dioxygenase, has been reported to be a flavin-containing enzyme[61]. With exception of these three enzymes, most of the other ring cleaving dioxygenases, if not all of them, contain nonheme iron as the sole cofactor. Among these, some enzymes contain the ferrous form of iron and some the ferric form[13, 81]. In the so-called double hydroxylation reactions, an electron transport system appears to be involved[62, 82, 83]. All the α-ketoglutarate-requiring dioxygenases that have so far been characterized require Fe(II) and ascorbic acid[14, 84]. The details concerning these enzymes will be described in a later section.

III. Nonheme Iron-Containing Dioxygenases

Of about 40 dioxygenases known up to date, more than 80% have iron built into their structure or require added iron for full activity. Cleavage of the benzene rings is a function that appears to depend almost entirely on nonheme iron-containing dioxygenases.

M. Nozaki

A) Catechol Dioxygenase

When a benzene ring is cleaved by a dioxygenase reaction, hydroxylation of the benzene ring usually proceeds to form catechol or phenol derivatives. When catechol derivatives are cleaved by the action of individual dioxygenases, three modes of ring fission have been demonstrated by microbial enzymes[85]:

(a) Cleavage of the bond between carbon atoms 3 and 4 to form muconic acid derivatives (intradiol cleavage), and
(b) that of carbon atoms 2 and 3 (proximal extradiol cleavage) and
(c) carbon atoms 4 and 5 (distal extradiol cleavage) to form α-hydroxy δ-substituted muconic semialdehyde and α-hydroxy δ-substituted muconic semialdehyde, respectively.

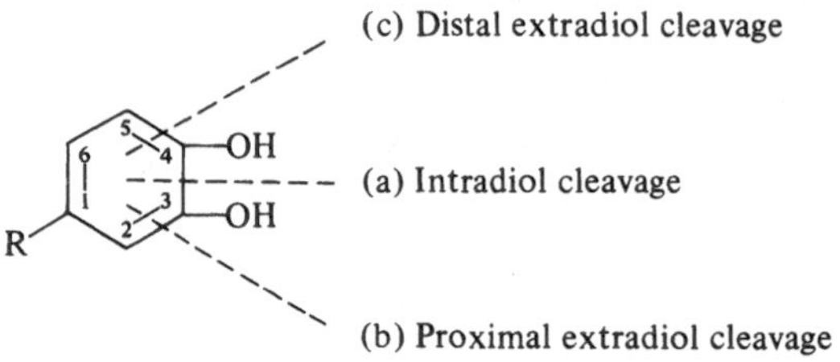

(c) Distal extradiol cleavage

(a) Intradiol cleavage

(b) Proximal extradiol cleavage

It is interesting to note that dioxygenases catalyzing the intradiol cleavage contain the ferric form of iron as the sole cofactor, whereas those catalyzing the extradiol cleavage contain the ferrous form.

Pyrocatechase[3] and protocatechuate 3,4-dioxygenase[86], both of which are typical examples of the intradiol cleavage, catalyze the reaction shown in Eqs. (2) and (12), respectively.

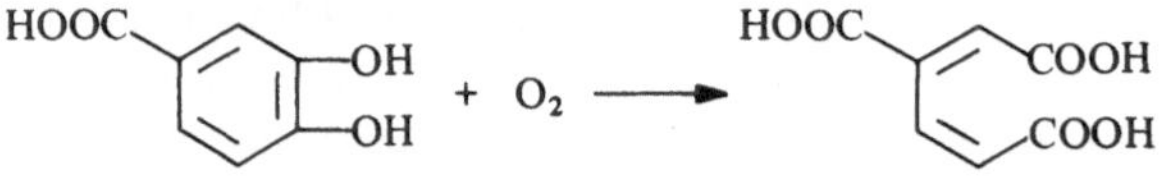

$$\text{(12)}$$

Both enzymes contain the ferric form of iron as a sole cofactor and show a characteristic absorption spectrum in the visible range[13, 81].

On the other hand, metapyrocatechase that catalyzes the reaction shown in Eq. (13) is a typical example of the extradiol enzymes[87, 88].

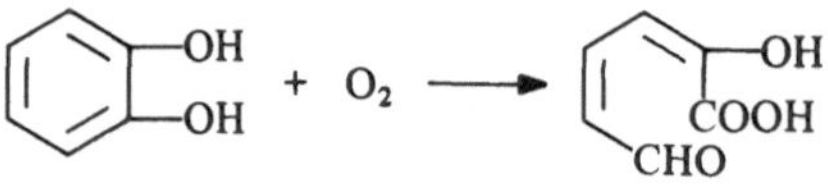

$$\text{(13)}$$

This enzyme is colorless and contains the ferrous form of iron as the sole cofactor[53, 54]. 3,4-Dihydroxyphenylacetate 2,3-dioxygenase catalyzes the reaction shown in Eq. (14) and is an example of the proximal extradiol cleavage[89–91]. Protocatechuate 4,5-dioxygenase catalyzes the reaction shown in Eq. (15) and is an example of the distal extradiol cleavage[92–94].

$$(14)$$

$$(15)$$

These two enzymes are also colorless and contain the ferrous form of iron[90, 91, 94].

The properties of catechol dioxygenases are summarized in Table 3.

1) Intradiol Dioxygenases

a. Catechol 1,2-Dioxygenase (Pyrocatechase)

Molecular Properties. Pyrocatechase from *Pseudomonas arvilla* C-1 was reported to have a molecular weight of 90,000 and to contain 2 g-atoms of ferric iron per mole of enzyme[55, 56]. However, more careful analyses have recently revealed that the enzyme actually has a molecular weight of approximately 60,000 and contain one atom of ferric iron[97]. The purified enzyme gives a single band on acrylamide gel electrophoresis. However, when the enzyme is treated with sodium dodecyl sulfate (SDS), it dissociates into two non-identical subunits α and β with molecular weights of 30,000 and 32,000 respectively[97]. The NH_2-terminal amino acid residues of α and β subunits are both found to be threonine, whereas the COOH-terminal residues are alanine and glycine, respectively[111]. Therefore, the enzyme consists of two non-identical subunits (α, β). One g-atom of ferric iron is present per pair of the subunits to form one molecule of the enzyme, which has a single active site[97].

Similar but not identical pyrocatechases have been purified to homogeneity from various microorganisms including *Brevibacterium fuscum* P-13[112, 113] and *Acinetobactor calcoaceticus*[114].

Brevibacterium pyrocatechase has a molecular weight of 64,000 and contains 1 g-atom of ferric iron per mole of enzyme. *Acinetobacter* pyrocatechase contains 2 g-atoms of iron per mole of enzyme, based on a molecular weight of 81,000. The subunit size determined by sodium dodecyl sulfate-gel electrophoresis is 40,000. The amino-terminal amino acid is methionine[114]. The substrate specificities of these two enzymes are somewhat different from that of *Pseudomonas* pyrocatechase (Table 4).

Spectral Properties. The absorption spectra of these three pyrocatechase are very similar to each other. A concentrated solution of the enzyme shows a distinct red color with a broad absorption between 390 and 650 nm. The peak is at about 440 nm and the molecular absorbance at 440 nm is estimated to be 4,670[55]. The trivalent iron bound to the enzyme is responsible for the visible absorption since the apo-enzyme shows neither significant absorption in the visible region nor enzyme activity which is restored upon reconstitution[56]. The visible absorption also de-

Table 3. Properties of catechol dioxygenases

	Catechol 1,2-dioxygenase	Protocatechuate 3,4-dioxygenase	Catechol 2,3-dioxygenase
Other names	Pyrocatechase	Protocatechuate oxygenase	Metapyrocatechase
Systematic name	Catechol:oxygen 1,2-oxidoreductase (decycling)	Protocatechuate:oxygen 3,4-oxidoreductase (decycling)	Catechol:oxygen 2,3-oxidoreductase (decycling)
EC number	1.13.11.1	1.13.11.13	1.13.11.2
Source material	*Pseudomonas arvilla* C-1	*Pseudomonas aeruginosa*	*Pseudomonas arvilla*
Substrate specificity (relative activity in parenthesis)	Catechol (100) 4-Methylcatechol (90) 3-Methylcatechol (8) 4-Chlorocatechol (3.6) 3-Methoxycatechol (0.8) Pyrogallol (0.6) Protocatechualdehyde (0.02)	Protocatechuic acid (100) Pyrogallol (2.4) Catechol (0.4) 3-Methylcatechol (0.4) 4-Methylcatechol (0.2) 3,4-Dihydroxyphenyl-acetic acid (0.2) 3,4-Dihydroxyphenyl-mandelic acid (0.1)	Catechol (100) 4-Methylcatechol (100) 3-Methylcatechol (62) 4-Chlorocatechol (51) Pyrogallol (33) Protocatechualdehyde (21) Protocatechuic acid (0.15)
Type of cleavage	Intradiol	Intradiol	Extradiol (proximal)
Molecular weight	6×10^4	70×10^4	14×10^4
Subunit composition	$\alpha\beta$	$(\alpha_2\beta_2)_8$	α_4
Molecular weight of subunits	$\alpha = 30,000$ $\beta = 32,000$	$\alpha = 22,500$ $\beta = 25,000$	$\alpha = 35,000$
Valence of Fe	Fe(III)	Fe(III)	Fe(II)
Fe content (g atoms/mole of enzyme)	1	8	4
Substrate bound (moles/mole of enzyme)	1	8	4
Color of enzyme	Red	Red	Colorless
Absorption peak (Absorption of a 1% solution)	280(8.93), 440(0.519)	280(13.2), 450(0.372)	280 (13.2)
ESR signal	$g = 4.28$	$g = 4.31$	No ESR signal

	3,4-Dihydroxyphenyl-acetate 2,3-dioxygenase	Protocatechuate 4,5-dioxygenase	3,4-Dihydroxy-9,10-secoandrosta-1,3,5(10)-triene-9,17-dione 4,5-dioxygenase
CD bands	Negative 222 nm Positive 250–300 nm Negative 327 nm, 500 nm	Negative 200–250 nm Positive 250–300 nm Negative 330 nm, 480 nm	Negative 225 nm Positive 250–300 nm
Turnover number	1,776	43,120	38,500
Specific activity (µmoles/min/mg protein)	29.6	61.6	275
Optimal pH	7.5	8.0	6.5
K_m values (µM) for substrate	4.0	30	3.0
for oxygen	20	43	7.0
Ref.	55, 56, 95–97)	98–105)	53, 54, 85, 106)
Other names	Homoprotocatechuate oxygenase	Protocatechuate 4,5-oxygenase	Steroid 4,5-dioxygenase 3-Alkylcatechol 2,3-dioxygenase
Systematic name	3,4-Dihydroxyphenylacetate: oxygen 2,3-oxidoreductase (decycling)	Protocatechuate:oxygen 4,5-oxidoreductase (decycling)	3,4-Dihydroxy-9,10-secoandrosta-1,3,5(10)-triene-9,17-dione: oxygen 4,5-oxidoreductase (decycling)
EC number	1.13.11.7.	1.13.11.8	1.13.11.25
Source material	*Pseudomonas ovalis*	*Pseudomonas* sp.	*Nocardia restricuts*
Substrate specificity relative activity in parenthesis	3,4-Dihydroxyphenylacetic acid (100) 3,4-Dehydroxyphenylpropionic acid (2.9) N-Formyl dopa (0.9) Dopa (0.3) Protocatechuic acid (0.13)	Protocatechuic acid (100)	3,4-Dihydroxy-9,10-secoandrosta-1,3,5(10)-triene-9,17-dione 3-Isopropylcatechol (100) 3-t-Butyl-5-methylcatechol (82.4) 3-Methylcatechol (7.4) 4-Methylcatechol (4.3) Catechol (0.6)

155

Table 3. (continued)

	3,4-Dihydroxyphenyl-acetate 2,3-dioxygenase	Protocatechuate 4,5-dioxygenase	3,4-Dihydroxy-9,10-secoandrosta-1,3,5(10)-triene-9,17-dione-4,5-dioxygenase
Type of cleavage	Proximal extradiol	Distal extradiol	Proximal extradiol
Molecular weight	14×10^4	15×10^4	28×10^4
Subunit composition	α_4	–	–
Molecular weight of subunits	$\alpha = 35{,}000$	–	–
Valence of Fe	Fe(II)	Fe(II)	Fe(II)
Fe content (g atoms/mole of enzyme)	4–5	1	1.13
Substrate bound (moles/mole of enzyme)	–	–	–
Color of enzyme	Colorless	Colorless	Colorless
Absorption peak (Absorption of a 1% solution)	280 (9.47)	280 (8.78)	280 (9.3)
ESR signal	No ESR signal	–	–
CD bands	Negative 225 nm Positive 260–310 nm	–	–
Turnover number	14,140	23,550	6,160
Specific activity (μmoles/min/mg protein)	101	157	22
Optimal pH	8.0	7.0	5.8–7.5
Km values (μM) for substrate	26	80	370 (3-isopropylcatechol)
for oxygen	–	54	180
Ref.	90, 91, 107, 108)	94)	109, 110)

Table 4. Substrate specificity of pyrocatechases from *Pseudomonas*, *Brevibacterium* and *Acinetobactor*

Substrate	Relative activity (%) of pyrocatechase from		
	Pseudomonas	*Brevibacterium*	*Acinetobactor*
Catechol	100	100	100
4-Methylcatechol	90	95	18
3-Methylcatechol	8	140–150	12
4-Chlorocatechol	3.6	–	–
Pyrogallol	0.6	85	0
4-Hydroxycatechol	–	30–50	–
4-Methoxycatechol	–	85	–
3-Isopropylcatechol	–	–	30

creases on reduction by the addition of sodium dithionite under anaerobic conditions and reappears on reoxidation by shaking the solution in air[56].

By the addition of the substrate, catechol, under anaerobic conditions, the color of the enzyme solution changes to greyish blue with a concomitant increases in absorbance at around 710 nm, indicating possible formation of an enzyme-substrate complex. This change in absorption spectrum is restored to the original level after catechol is degraded to *cis,cis*-muconic acid by the addition of oxygen[55].

Electron-Spin Resonance Properties. A concentrated solution of pyrocatechase shows a sharp ESR signal at g = 4.28, known to be due to the high spin state of ferric iron[56]. The signal markedly decreases upon the addition of sodium dithionite anaerobically and is restored when the solution is exposed to air. The signal also decreases with a concomitant loss of the enzyme activity under a variety of conditions. The apoenzyme shows no ESR signal. The signal disappears instantaneously when the substrate, catechol, is added to the enzyme under anaerobic conditions and is restored to the original level after all of the substrate is degraded to *cis,cis*-muconic acid by the addition of air[56, 115]. A similar signal change has also been reported with *Brevibacterium* pyrocatechase[107, 113, 116]. The disappearance of the signal at g = 4.28 by the anaerobic addition of catechol may be associated with the formation of an enzymatically active enzyme-substrate complex, since similar changes are observed upon the addition of all other substrates. However, the disappearance of the signal does not necessary mean that the iron in the enzyme is reduced since magnetic susceptibility experiments indicate that the iron in the enzyme-substrate complex is still in the high spin ferric state[56].

Circular Dichroism. The circular dichroism (CD) bands of *Pseudomonas* pyrocatechase can be classified into three groups:
1) negative bands at 327 and 500 nm;
2) positive bands at 250 to 300 nm; and
3) a negative band at 222 nm characteristic of an α-helix in proteins[95].

The negative bands at 327 and 500 nm seem to be the result of chelation between trivalent iron and amino acid residues of the protein, since the bands disappear upon either removal or reduction of the iron. Upon the addition of the substrate under anaerobic conditions, the band at 500 nm shifts to 540 nm and the band at 327 nm is diminished. A part of the positive bands between 250 and 300 nm is also associated with the enzyme activity, since they are partially diminished upon removal of bound iron[95]. Similar but not identical CD bands have been reported for *Brebacterium* pyrocatechase[113].

Serological Properties. Antisera prepared against purified preparation of *Acinetobactor* pyrocatechase cross-react and inhibit the enzyme activity in the crude extracts prepared from other strains of *Acinetobactor calcoaceticus*, but fail to cross-react and inhibit isofunctional enzymes prepared from strains of *Pseudomonas*, *Nocardia*, and *Alcaligenes*[114].

b. Protocatechuate 3,4-Dioxygenase

Molecular Properties. Protocatechuate 3,4-dioxygenase has been obtained in crystalline form from p-hydroxybenzoate-induced cells of *Pseudomonas aeruginosa*[98]. The enzyme has a molecular weight of approximately 700,000, and contains about 8 g-atoms of ferric iron and eight substrate binding sites per mole of enzyme. From the fact that the enzyme is dissociated into homogeneous subunits of approximately. 90,000 daltons by an alkaline treatment, in accordance with an electron microscopic observation, the enzyme was reported to consist of eight identical subunits, each of which appears to contain one atom of ferric iron and a substrate binding site[98, 99]. However, this 90,000 dalton subunit has recently been found to dissociate further into four smaller subunits of two non-identical types ($\alpha_2\beta_2$), upon sodium dodecyl sulfate gel electrophoresis[105]. These two subunits can be separated by SP-Sephadex chromatography in the presence of urea at pH 5.8. The molecular weights of the α and β subunits are estimated to be 22,500 and 25,000, respectively. Isoelectric focusing of the enzyme reveals that the isoelectric points of the α and β subunits are 5.2 and 9.5, respectively[105].

The amino acid compositions of each subunit and the native enzyme are shown in Table 5. There are distinct differences between the two subunits. As would be expected from the isoelectric points, the α subunit contains more acidic amino acids than the β subunit does, and the β subunit contains more basic amino acids than the α subunit does. The amino acid composition of the native enzyme obtained by actual analyses agrees quite well with that calculated from the compositions of the α and β subunits, based on the assumed subunit structure of $(\alpha_2\beta_2)_8$. These results confirms the assumption that the native enzyme consists of 8 protomers, each of which is composed of a pair of two non-identical subunits $(\alpha_2\beta_2)$[97, 105].

The NH_2-terminal residue of the native enzyme, as well as those of the α and β subunits are found to be proline. The COOH-terminal amino acid for both the native enzyme and the α subunit is found to be phenylalanine. However, the COOH-terminal residue of the β subunit is still unknown. The NH_2-terminal sequences of these two subunits already determined are as follows[105].

α subunits: Pro-Ile-Glu-Leu-Leu-Pro-Glu-Thr-Pro-Ser-Glx-Thr-Ala-Gly-
β subunits: Pro-Ala-Gln-Asp-Asn-Ala-Arg-Phe-Val-Ile-Arg-Asx-Arg-Asx-
Except for the NH_2-terminal residue, the amino acids in each degradation step of the two subunits are different, suggesting that the enzyme is composed of two non-identical polypeptide chains. These results suggest that the enzyme consists of eight

Table 5. Amino acid compositions of native protocatechuate 3,4-dioxygenase and its subunits[a]

	α	β	$(\alpha_2 + \beta_2)X8$	Native[b]
Lys	5	8	208	182
His	5	7	192	194
Arg	11	14	400	351
Asp	24	23	752	718
Thr	10	9	304	286
Ser	5	9	224	205
Glu	24	16	640	513
Pro	13	21	544	477
Gly	14	18	512	460
Ala	19	15	544	477
Val	11	9	320	265
Met	1	2	48	46
Ile	12	15	432	364
Leu	19	17	576	459
Tyr	7	8	240	188
Phe	9	9	288	247
1/2 Cys	2	4	96	95
Trp	6	4	160	160

[a] The values are expressed as number of residues on the basis of the assumption that the native enzyme has a molecular weight of 700,000 daltons; the α subunit, 22,500 daltons; and the β subunit, 25,000 daltons.

[b] Data were taken from Ref.[98].

identical protomers, each containing two pairs of two non-identical subunits, $(\alpha_2\beta_2)_8$. This protomer $(\alpha_2\beta_2)$ appears to contain one atom of ferric iron, forming one active site of the enzyme[97, 105].

Another protocatechuate 3,4-dioxygenase has been purified to homogeneity from p-hydroxybenzoate-induced cells of *Acinetobacter calcoaceticus*[117]. The enzymes from *Acinetobacter* and *Pseudomonas* are quite similar in their molecular weight, molecular size, and iron content. However, the specific activity of the *Acinetobacter* enzyme is about one-third of that of the *Pseudomonas* enzyme, despite of their similar iron contents[117].

Spectral Properties. Like another intradiol dioxygenase, pyrocatechase, protocatechuate 3,4-dioxygenase has a deep red color with a broad absorption between 400 and 650 nm. The red color disappears upon the addition of sodium dithionite under anaerobic conditions and reappears when the solution is exposed to air or when potassium ferricyanide is added to the solution under anaerobic conditions[99]. The apoenzyme prepared by prolonged anaerobic dialysis against a buffer solution containing both *o*-phenanthroline and sodium dithionite shows no visible absorption. The visible absorption spectrum characteristic of the native enzyme reappears upon reconstitution[101]. These results suggest that the visible absorption is attributable to the trivalent iron bound to the enzyme.

The visible absorption spectrum of the enzyme exhibits an increase in absorbance with a slight red shift of the peak by the addition of the substrate under anaerobic

conditions, indicating the possible formation of an enzyme-substrate complex[98–100].
The spectrum is restored to the original one after the substrate is exhausted by the
addition of oxygen. The spectrum of the enzyme-substrate complex decreases
markedly when sodium dithionite is added to the complex and is restored to that of
the complex upon the addition of ferricyanide[99]. Further addition of oxygen con-
verts the spectrum to that of the original enzyme. These results suggest that the iron
in the enzyme-substrate complex is also in the trivalent state.

When the enzyme-substrate complex is mixed with a buffer solution containing
oxygen, the absorbance at 470 nm decreases rapidly and after 3 msec reachs the
minimum and increases again rather gradually returning to the original level. Series
of similar experiments at different wavelengths give a short-lived new spectral species
reconstructed from the changes in absorbance at each wavelength and the spectrum
of the enzyme-substrate complex[118]. This new spectral species is characterized by
a broad absorption band with a maximum between 500 and 520 nm, distinct from
those of the enzyme or the enzyme-substrate complex, and is observed only in the
presence of both protocatechuic acid and molecular oxygen[118]. A spectrum similar
to the one observed with protocatechuic acid is also observed with substrate analogues such as 3,4-dihydrophenylacetic acid and 3,4-dihydroxyphenylpropionic
acid during the steady state[100].

The rate constant for the decomposition of the complex is calculated to be
94 sec^{-1} which agrees quite well with the turnover number of the enzyme (95 sec^{-1}
per active site of the enzyme) as determined from the overall reaction[118]. Detailed
kinetic analyses of the enzyme with 3,4-dihydroxyphenylacetic acid and 3,4-dihydroxy-
phenylpropionic acid as the substrate reveal that the new spectral species is really
an obligatory intermediate and is a ternary complex of oxygen, substrate and enzyme,
i.e. an oxygenated intermediate[100]. In order to form the oxygenated intermediate,
the presence of an organic substrate is necessary. Therefore, it is reasonable to
assume that the organic substrate combines with the enzyme first and then reacts
with oxygen to form the oxygenated intermediate [Eq. (16)].

$$E \underset{K_{-1}}{\overset{k_{+1}S}{\rightleftharpoons}} ES \underset{K_{-2}}{\overset{k_{+1}O_2}{\rightleftharpoons}} ESO_2 \overset{k_{+3}}{\longrightarrow} E + P \tag{16}$$

If the enzymic reaction proceeds via the mechanism given by Eq. (16), the Km value
for the organic substrate and the oxygen should be expressed by k_{+3}/k_{+1} and
$(k_{-2} + k_{+3})/k_{+2}$, respectively. The values of k_{+3} are in good agreement with the
turnover numbers based on the active site of the enzyme, indicating that the decomposition of the oxygenated intermediate is the rate-limiting step for the overall reaction. The values of k_{+3}/k_{+1} and $k_{-2} + k_{+3}/k_{+2}$ also coincide roughly with the Km
values for the organic substrate and oxygen, respectively[100].

Electron Spin Resonance and Mossbauer Properties. Protocatechuate 3,4-dioxy-
genase shows a sharp ESR signal at g = 4.31, typical of high-spin ferric iron in a
"rhombic" environment. The signal height diminishes instantaneously upon the addition of the substrate under anaerobic conditions. The decreased signal is restored to
that of the original level when the substrate is completely converted to the product
upon the introduction of air[99]. By measuring the temperature dependence of the

low temperature ESR signal, Peisach *et al.* suggest that an iron ligand environment in the enzyme is similar to ferric rubredoxin, *i.e.*, a tetrahedral arrangement of four cysteinyl sulfurs[119, 120]. Upon the addition of the substrate or substrate analogs under anaerobic conditions, new resonance at g = 6.4 and 5.6 are observed which result from Fe(III) in a nearly tetragonal environment. The situation is analogous to the geometric environment of ligands that exist in heme, suggesting that the binding of the substrate to the enzyme causes changes in ligand symmetry, which makes the iron accessible for O_2 binding[119, 120]. On the other hand, from the results of a combined ESR and Mössbauer investigation of ^{57}Fe-enriched protocatechuate 3,4-dioxygenase, Que, Jr., *et al.* suggest that the iron coordination in the enzyme differs markedly from rubredoxin and that the iron atoms are in a previously unrecognized environment[104]. The Mössbauer and ESR data on the ternary complex of the enzyme with 3,4-dihydroxphenylpropionate and O_2 establish the iron atoms to be in a high-spin ferric state characterized by a large and negative zero-field splitting, $D = \simeq -2 \ cm^{-1}$ [104].

Circular Dichroism. The native enzyme from *Pseudomonas* exhibits several positive CD bands between 250 and 300 nm and two, low-intensity, negative bands at 330 and 480 nm. In the presence of protocatechuic acid and the absence of O_2, spectral changes are evident in the side chain and visible regions. There is a shift in the aromatic-region maximum from 275 to 267 and in the visible region from 330 to 348 and from 480 to 555 nm. No spectral changes are observed upon the removal or addition of only O_2 [102]. The apo- and holo-enzymes show identical spectra in the ultraviolet region between 200 and 250 nm (peptide backbone region), but the low intensity negative bands at 330 and 480 nm of the holoenzyme are completely absent in the apoenzyme. Either with or without the iron, the enzyme protein binds protocatechuate and has a greater positive circular dichroism increase at 240–260 nm[103].

The circular dichroism spectra of *Acinetobactor* protocatechuate 3,4-dioxygenase are similar but somewhat different from those of the *Pseudomonas* enzyme, indicating some conformational differences between these two enzymes[117]. However, it is interesting to note that CD spectra of protocatechuate 3,4-dioxygenase and pyrocatechase, another Fe(III)-containing enzyme, are in general very similar to each other, in spite of their slight differences.

Serological Properties. Antisera prepared against protocatechuate 3,4-dioxygenase from *Pseudomonas aeruginosa* cross-react homologously with the isofunctional enzyme from organisms of the genus *Pseudomonas* but do not cross-react with isofunctional enzymes from other genera including *Acinetobactor calcoacelicus* and *Arthrobacter pascens*[121].

2) Extradiol Dioxygenases

a. Catechol 2,3-Dioxygenase (Metapyrocatechase)

General Properties. Metapyrocatechase was the first dioxygenase obtained in a crystalline form[53]. The molecular properties of this enzyme are summarized in Table 3. The enzyme is extremely unstable in the presence of air[88]. However, low concentrations of organic solvents such as acetone and ethanol protect the enzyme from inactivation. Hence, the crystallization of metapyrocatechase was achieved using buffer solutions containing 10% acetone in all purification procedures[53]. Iron is the sole cofactor of the enzyme and seems to be of the ferrous form[54].

Unlike intradiol dioxygenases, metapyrocatechase is colorless and shows neither significant absorption in the visible range nor any ESR signal at around g = 4.3. Therefore, spectral analyses of the enzyme has been hampered and spectrally distinct reaction intermediates like the ones demonstrated with intradiol dioxygenases are not observed. However, steady state kinetic analyses with metapyrocatechase reveal that a ternary complex of the enzyme with oxygen and the organic substrate is involved in the reaction[122]. Furthermore, dead-end inhibitors such as o-nitrophenol and m-phenanthroline inhibit the enzyme competitively with respect to the organic substrate, catechol, and non-competitively with respect to the other substrate, oxygen. These results are consistent with an ordered bi-uni mechanism where the organic substrate (S) first combines with the enzyme (E) and then reacts with oxygen to form a ternary complex (ESO_2) [Eq. (17)]. This reaction sequence is quite similar to that of protocatechuate 3,4-dioxygenase[100].

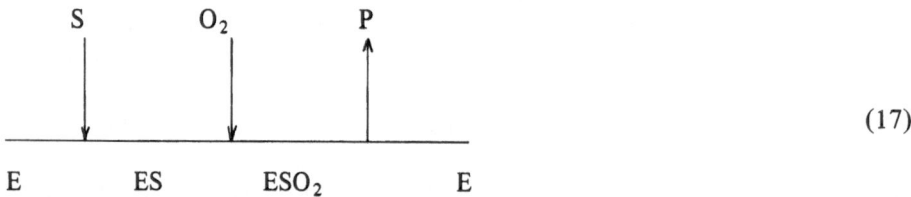

$$(17)$$

CIRCULAR DICHROISM. Metapyrocatechase shows a strong negative CD band at 225 nm and positive dichroic bands between 250 and 300 nm. No significant CD bands are observed in the visible range[106]. However, upon the addition of the substrate, catechol, under unaerobic conditions, a new negative band appears at 317 nm, accompanied by a decrease in the band at 300 nm. Although apoenzyme can combine with catechol[123], no such changes in CD is observed with the apoenzyme, indicating that the iron is involved in the negative CD band at 317 nm[106].

b. 3,4-Dihydroxyphenylacetate 2,3-Dioxygenase

This enzyme has also been obtained in a crystalline form from extracts of *Pseudomonas ovalis* grown with p-hydroxyphenylacetate as the major carbon source[90]. Like metapyrocatechase, the enzyme is colorless and shows no ESR signal in the native state. An ESR signal at g = 4.3 appears in the presence of the substrate and oxygen, and disappears after the exhaustion of oxygen. From these results, a valence change of the iron during the catalysis and the involvement of a ternary complex, enzyme (Fe III)-substrate-oxygen, as an intermediate are suggested[107]. However, no kinetic analysis has been carried out to show whether or not the complex is an obligatory intermediate in the reaction.

The *Pseudomonas* enzyme shows a strong negative CD band at 225 nm (peptide backbone region) and small positive bands between 260–310 nm. However, no CD band was observed in the visible range[91].

Another 3,4-dihydroxyphenylacetate 2,3-dioxygenase has been purified from the thermophilic organism *Bacillus stearothermophilus*, grown with 4-hydroxyphenylacetic acid[124]. The enzyme appears to be homogeneous as judged by disc-gel electrophoresis and sedimentation equilibrium measurements. The molecular weight is determined to be 106,000. The enzyme is further dissociated in sodium dodecyl sulfate into 33,000 to 35,000 dalton subunits, suggesting that the enzyme consists of 3 identical subunits. The enzyme is fairly stable on heating and shows maximal activity at about 57 °C. However, the specific activity of the *Bacillus* enzyme is much lower and less than one hundredth of that of the *Pseudomonas* enzyme[90].

c. Protocatechuate 4,5-Dioxygenase

As shown in Table 3 an almost homogeneous preparation of protocatechuate 4,5-dioxygenase purified from a *Pseudomonas sp.* grown with protocatechuic acid as the major carbon source has a specific activity of 160 μmoles/min/mg of protein with a molecular weight of approximately 150,000. One g-atom of ferrous iron is present per mole of enzyme[94]. The enzyme has also been purified from extracts of *p*-hydroxybenzoate-induced *Pseudomonas testeroni*[125]. The enzyme has a molecular weight of about 140,000 with a specific activity of about 12.

Zabinski *et al.*[126] have also reported that the enzyme purified from *Pseudomonas testosteroni* contains four iron atoms per mole of enzyme. Substrate specificity studies of the enzyme reveal that the carboxyl group and the hydroxyl groups are essential for cleavage of the aromatic ring. However, substitution at C_5 of the benzene ring is allowed because gallic acid and 5-methoxygallic acid function as alternative substrates for protocatechuic acid. The iron in the enzyme is not detectable by ESR. However, upon incubation of the enzyme with the substrates protocatechuic acid, gallic acid, and 5-methoxygallic acid, respectively, the iron becomes high-spin ferric. This change at the active site is independent of the presence of oxygen, but absolutely dependent on the addition of each respective catechol substrate. The enzyme-substrate complex shows an ESR signal at g = 4.3. This signal disappears when the enzyme has converted the substrate to product. Preincubation of the enzyme with the competitive inhibitors, protocatechualdehyde or 4-nitrocatechol, also gives an ESR signal at g = 4.3. Mössbauer data show that all four iron atoms in the enzyme-inhibitor complex are high-spin ferric and equivalent, and that the iron atoms in the native enzyme are either in a low-spin ferrous state or the enzyme has two active sites each containing two antiferro-magnetically-coupled high-spin ferric irons[126].

d. Steroid 4,5-Dioxygenase

2,4-Dihydroxy-9,10-secoandrosta-1,3,5(10)-triene-9,17-dione-4,5-dioxygenase (steroid 4,5-dioxygenase), catalyzes the reaction shown in Eq. (18)[109].

$$+ \ O_2 \ \longrightarrow \qquad\qquad\qquad\qquad (18)$$

The enzyme was purified to a state of apparent homogeneity from *Nocardia restrictus*. The general properties of the enzyme are summarized in Table 3.

The enzyme gives intersecting initial velocity plots that conform to a sequential mechanism. Experiments with 4-isopropylcatechol, a dead-end inhibitor and a structural analog of the organic substrate, reveals that the enzyme is inhibited by the inhibitor competively with respect to the organic substrate and uncompetitively with respect to molecular oxygen. These results are consistent with an ordered Bi-Uni mechanism where molecular oxygen is added first, followed by organic substrate, and the product is then released[110]. This mechanism is different from those of protocatechuate 3,4-dioxygenase and metapyrocatechase[122].

B) Other Nonheme Iron-Containing Dioxygenases

The nonheme iron-containing dioxygenases other than the catechol dioxygenases mentioned above are listed in Table 6. For the details of these enzymes, the readers are referred to the original references of individual enzymes or the review articles[14,81].

C) Extradiol Cleavage of Catechols by Intradiol Dioxygenases

As mentioned previously, it has been believed that the site of cleavage of an aromatic ring is strictly specific for each enzyme, namely, the ferric iron-containing dioxygenases would exclusively cleave the catechol ring in the intradiol manner, whereas the ferrous iron-containing dioxygenases would cleave it in the extradiol manner. However, a ferric iron-containing dioxygenase, pyrocatechase, from *Pseudomonas arvilla C-1* is found to catalyze not only an intradiol cleavage but also a proximal extradiol cleavage when 3-substituted catechols are used as the substrate.

The enzyme catalyzes the intradiol cleavage of catechol with the insertion of 2 atoms of molecular oxygen to form *cis,cis*-muconic acid. The enzyme also catalyzes the oxidation of various catechol derivatives, including 4-methylcatechol, 4-chlorocatechol, 4-formylcatechol (protocatechualdehyde), 4,5-dichlorocatechol, 3,5-dichlorocatechol, 3-methylcatechol, 3-methoxycatechol, and 3-hydroxycatechol (pyrogallol). All of these substrates give products having an absorption maximum at around 260 nm characteristic of *cis-cis*-muconic acid derivatives. However, when 3-methylcatechol is used as a substrate, the product formed shows an absorption maximum at 390 nm besides that at 260 nm. These two absorption maxima are found to be attributable to two different products, 2-hydroxy-6-oxo-2,4-heptadienoic acid (*1*), and 5-carboxy-2-methyl-2,4-pentadienoic acid (2-methylmuconic acid), (*2*) [Eq. (19)][96].

$$(19)$$

Compound *1* is produced by extradiol cleavage of the bond between the carbon atom carrying the hydroxyl group and the carbon atom carrying the methyl group, and compound *2* by intradiol cleavage of the bond between two hydroxyl groups. Similarly, 3-methoxycatechol gives two products: 2-hydroxy-5-methoxycarbonyl-2,4-pentadienoic acid and 5-carboxy-2-methoxy-2,4-pentadienoic acid (2-methoxy-muconic acid)[96].

With 3-methylcatechol as substrate, the ratio of intradiol and extradiol cleavage activities of *Pseudomonas* pyrocatechase during purification is almost constant and is about 17. The final preparation of the enzyme is almost homogeneous and all attempts to resolve the enzyme into two components with separate activities, including inactivation of the enzyme with urea or heat, treatment with sulfhydryl-blocking reagents or chelating agents, and inhibition of the enzyme with various inhibitors, are unsuccessful. These results suggest that *Pseudomonas* pyrocatechase catalyzes simultaneously both intradiol and extradiol cleavages of some 3-substituted catechols[96]. On the other hand, pyrocatechase purified from *Brevibacterium fuscum* catalyzes only the intradiol cleavage even with 3-substituted catechols as a substrate[96].

The extradiol cleavage of 3-methylcatechol by the intradiol enzymes is recently confirmed by Hou *et al.* [160] with pyrocatechases from various microorganisms. The isofunctional enzymes from species of *Acinetobactor*, *Pseudomonas*, *Nocardia*, *Alcaligens*, and *Corynebacterium* oxidize 3-methylcatechol according to both the intradiol and extradiol cleavage patters. However, the enzymes prepared from *Brevibacterium* and *Arthrobacter* have only the intradiol activity. The enzyme with no extradiol cleavage activity oxidize both 3-methylcatechol and 4-methyl-catechol at almost the same rate as catechol. Pyrocatechase from other microorganisms except *Acinetobactor calcoaceticus* ADP-96 oxidizes 4-methylcatechol at a rate about one order greater than 3-methyl-catechol[160].

IV. Heme-Containing Dioxygenases

Dioxygenases that catalyze the cleavage of the indole ring of tryptophan derivatives are known to be the heme proteins, tryptophan 2,3-dioxygenase (tryptophan pyrro-

Table 6. Nonheme iron containing dioxygenases other than catechol dioxygenases listed in Table 3

Enzyme	EC Number	Reaction	Cofactor	Distribution	Ref.
3-Hydroxyanthranilate oxygenase	1.13.11.6		Fe(II)	Liver, kidney	127—132)
Homogentisate oxygenase	1.13.11.5		Fe(II)	Liver, Pseudomonas	133—141)
Gentisate 1,2-dioxygenase	1.13.11.4		Fe	Pseudomonas Moraxella	142—145)
3-Carboxyethylcatechol 2,3-dioxygenase	1.13.11.6		Fe	Achromobactor	146)
2,5-Dihydroxypyridine oxygenase	1.13.11.9		Fe[1]	Pseudomonas	147—149)

Enzyme	EC number	Reaction	Cofactor	Source	References
2,3-Dihydroxyindole 2,3-dioxygenase	1.13.11.23	2,3-dihydroxyindole $+ O_2 \rightarrow$ [intermediate] $\rightarrow$ anthranilic acid (COOH, NH_2) $+ CO_2$	Fe^{a}	Bacteria	150)
Ascorbate 2,3-dioxygenase	1.13.11.13	ascorbate $+ O_2 \rightarrow$ (COOH–COOH) $+$ (COOH, HOCH, HCOH, CH_2OH)	Fe	Myrothecium verrucaria	151)
Cysteamine oxygenase	1.13.11.9	(CH_2–SH, CH_2–NH_2) $+ O_2 \rightarrow$ (CH_2SO_2H, CH_2–NH_2)	Fe(III) sulfide	Various tissues of different animals	68–70) 152–157)
Cysteine oxygenase	1.13.11.20	(CH_2–SH, H–C–NH_2, COOH) $+ O_2 \rightarrow$ (CH_2–SO_2H, H–C–NH_2, COOH)	Fe^{a} NAD(PH)H	Liver	71–73) 158–159)
Lipoxygenase	1.13.11.12	CH_3–$(CH_2)_4$–CH=CH–CH_2–CH=CH–$(CH_2)_7$–COOH $+ O_2 \rightarrow$ CH_3–$(CH_2)_4$–CH–CH=CH–CH=CH–$(CH_2)_7$–COOH (OOH)	$Fe(III)^{b}$	Soybean	74–76)

[a] Iron is not detected in the enzyme preparation, but the enzyme is inhibited by iron chelators.

[b] Lipoxygenase purified from fungus contains protoporphyrine IX (see Section IV).

lase (EC 1.13.11.11) and indoleamine 2,3-dioxygenase. Recent review articles concerning these enzymes are available [13, 14, 161]. Recently, lipoxygenase from fungus has also been demonstrated to be a hemoprotein [162, 163].

A) Tryptophan 2,3-Dioxygenase

In 1936, Kotake and Masayama described the conversion of L-tryptophan to L-kynurenine *in vitro* by crude extracts of rabbit liver and named the enzyme involved in the cleavage of the pyrrole ring of tryptophan "tryptophan pyrrolase" [164]. The enzyme was then studied extensively by Knox and his co-workers [165]. An enzyme which has similar properties was found in the cells of a Pseudomonad which were grown in a medium containing tryptophan as a major carbon source [166]. By tracer experiments, the bacterial enzyme was shown to be a dioxygenase catalyzing the following reaction [Eq. (20)].

$$
\text{[indole-CH}_2\text{-CH(NH}_2\text{)-COOH]} + O_2 \longrightarrow \text{[product]} \tag{20}
$$

General Properties of Tryptophan 2,3-Dioxygenase. The enzyme has been purified either from cells of *Pseudomonas,* grown in the presence of tryptophan [58, 167] or from the livers of glucocorticoid and L-tryptophan-treated rats [168]. The enzymes from both sources appear to be soluble proteins and have been purified to almost homogeneous state [167, 168]. The molecule weight of the Pseudomonad and hepatic enzymes are determined to be 122,000 [167] and 167,000 [168], respectively. Dissociation of the native enzyme with sodium dodecyl sulfate yields subunits of molecular weights of 31,000 for the Pseudomonad and 43,000 for the hepatic enzymes [167, 168]. Thus, each of the enzymes are tetramers of subunits of the same size. Upon polyacrylamide gel electrophoresis at pH 12.5, the subunits of the hepatic enzyme separates into two distinct species, indicating that the hepatic enzyme is composed of two pairs of identical subunits of equivalent mass $(\alpha_2\beta_2)$ [168]. Whether the subunits of the Pseudomonad enzyme are identical or distinguishable into two distinct species remains uncertain at present time.

The enzymes of either microbial or mammalian origin are hemoproteins [57, 58, 167, 168]. The concentrated solutions of the enzyme show absorption spectra in the visible and the ultraviolet range characteristic of ferriheme protein [58, 167, 168]. The heme of the Pseudomonad enzyme is identified as protohematin IX and the maximal turnover number of the enzyme is found to be 1,200 min^{-1} [58]. Analyses of the heme concentration using the dipyridine hemochromogen are compatible in the presence of 2 moles of heme per tetramer of the Pseudomonad and hepatic enzymes [168, 169]. Besides the heme, the presence of 2 g-atoms of copper per mole of tetrameric enzyme and its involvement in the catalysis have been reported by Feigelson and his co-workers [169, 170]. On the other hand, Poillon *et al.* [167] and Ishimura and Hayaishi [171] found only minute quantities of copper in their preparations and regarded these as adventitious.

Spectral Properties. The purified Pseudomonad enzyme exhibits a typical spectrum of a high-spin protohemoprotein both in its ferric and ferrous states. The heme in the enzyme is autoxidizable and the enzyme is usually isolated in the ferric form[57, 58]. The ferric enzyme is catalytically inactive unless some reducing agents, such as H_2O_2 or ascorbic acid, are added, suggesting that the ferrous form of enzyme is catalytically active[58]. When the substrate tryptophan is added to either the ferric or ferrous state of the enzyme under anaerobic conditions, slight shifts of the Soret band to the longer wavelength with some hypochlomicity are observed. This observation seems to indicate that tryptophan combines with the enzyme irrespective of the valency state of the heme, forming an enzyme-substrate (ES) complex. When the ES complex of the ferrous form of the enzyme is mixed with molecular oxygen, a new spectral species appears immediately with absorption maxima at 418, 545, and 580 nm. When the oxygen is used up, the spectrum reverts almost instantaneously to the ferrous state; this process can be repeated until all the tryptophan in the system is consumed. Available evidence indicates that the observed new spectrum is due to a ternary complex of oxygen, the enzyme and L-tryptophan, and represents the oxygenated intermediate[58]. By means of rapid reaction spectrophotometry, the oxygenated form is shown to be an obligatory intermediate of the reaction. In order to form the oxygenated intermediate, the presence of the organic substrate is necessary. Thus a sequential order of substrate binding is suggested in which tryptophan combines with the enzyme first and then reacts with oxygen to form the ternary complex[58]. Steady-state kinetic analyses of the reaction are also consistent with a Bi-Uni ordered mechanism in which L-tryptophan binds before oxygen[172]. These situations are quite similar to those observed with the nonheme iron-containing dioxygenases mentioned previously.

On the other hand, Feigelson and his associates[169, 170, 172–175] have reported that not only heme but also copper is an essential cofactor in both Pseudomonad and hepatic enzymes and that there are three different oxidation-reduction states of the enzyme, namely, a fully reduced form, $E(Cu^+, Fe^{2+})$; a half-reduced form, $E(Cu^+, Fe^{3+})$ or its valence isomer $E(Cu^{2+}, Fe^{2+})$; and a fully-oxidized form, $E(Cu^{2+}, Fe^{3+})$. From the spectrophotometric analyses of the reaction and the experiments with various chelators, they have suggested that both the fully-reduced and the half-reduced forms of the enzyme are catalytically active whereas the fully-oxidized form is inactive. Steady-state kinetic analyses of these two active forms of the enzyme conform to the mechanism that the fully-reduced form combines with tryptophan first, followed by oxygen combination whereas with the half-reduced form, oxygen binds with the enzyme before tryptophan[172, 175]. These results suggest that it is not necessary for catalytically active enzyme to be in the ferrous state. This explanation is inconsistent with that proposed by Hayaishi and his associates[58, 178, 179]. They insist that the heme is the sole cofactor of the enzyme, since the purified preparation of the enzyme contains negligible amounts of copper[171, 176].

The purified Pseudomonad enzyme shows the ESR signal at g = 6, characteristic of a high-spin protohemoprotein. Upon the addition of the organic substrate tryptophan to the enzyme, the signal becomes narrow and sharp, indicating that the axial symmetry of the heme increases while the total spin concentration due to the heme iron is unchanged[177]. These results indicate that the organic substrate binds with the enzyme and alters the conformation of the protein. The fact that the presence of L-tryptophan enhances the reactivity of heme in the enzyme toward ligands such as oxygen, CO, and cyanide supports the above interpretation[178, 179].

B) Indoleamine 2,3-Dioxygenase

In 1957, Kotake and Ito found that rabbits fed D-tryptophan excreated D-kynurenine in urine and suggested the occurrence of an enzyme in rabbit tissue that catalyzes the oxidative ring cleavage of D-tryptophan[180]. An enzyme activity capable of oxidizing D-tryptophan to D-kynurenine was shown by using a homogenate of rabbit small intestine and the enzyme was tentatively designated as D-tryptophan pyrrolase[181]. Since the purified enzyme catalyzes the oxidation of not only D-tryptophan but also L-isomer, the enzyme is referred to as intestinal tryptophan 2,3-dioxygenase (pyrrolase)[182]. The purified preparation of the enzyme shows the absorption spectra characteristic of hemoprotein[182].

The enzyme requires methylene blue and ascorbic acid as cofactors[181, 182]. Xanthine oxidase with hypoxanthine can replace ascorbic acid but hydrogen peroxide or its generating system can not[182]. A highly purified preparation of the enzyme from rabbit intestine has a broad substrate specificity and is found to catalyze the oxygenative ring cleavage of 5-hydroxytryptophan, serotonin and tyramine besides the D- and L-isomers of tryptophan[183]. A similar enzyme partially purified from rabbit brain catalyzes the oxygenative cleavage of the pyrrole moiety of melatonin yielding N^γ-acetyl-N^2-formyl-5-methoxykynurenine in the presence of methylene blue and ascorbic acid[184]. The brain enzyme is also able to degrade tryptophan and 5-hydroxytryptophan as well as serotonin. Therefore the term "indoleamine 2,3-dioxygenase" is proposed to designate these enzymes[184].

The intestinal enzyme is inhibited by the presence of a highly purified preparation of superoxide dismutase obtained either from bovine erythrocytes or green peas[185]. Furthermore, when the superoxide anion (O_2^-) produced by electrolytic reduction of molecular oxygen or potassium superoxide (KO_2) is added to the reaction mixture in the absence of ascorbic acid, the intestinal enzyme catalyzes the conversion of tyrptophan to formylkynurenine[185, 186]. These results suggest the participation of O_2^- in the catalytic process of indoleamine 2,3-dioxygenase. On the other hand, since tetrahydrobiopterine and 2-amino-4-hydroxy-6,7-dimethyl-5,6,7,8-tetrahydropteridine ($DMPH_4$) are found to accelerate the enzymic reaction in the presence of catalase even in the absence of methylene blue, it is suggested that tetrahydropteridine may be the natural cofactor of indoleamine 2,3-dioxygenase[187]. The reaction stimulated by tetrahydrobiopterin alone is inhibited by superoxide dismutase. In contrast, the reaction stimulated by tetrahydrobiopterin plus methylene blue is hardly inhibited at all by superoxide dismutase[187]. On the other hand, experiments with ^{18}O-labeled potassium superoxide ($K^{18}O_2$) and molecular oxygen ($^{18}O_2$) reveal that O_2^- is utilized and incorporated into the products by indoleamine 2,3-dioxygenase and that molecular oxygen is also utilized but to a lesser extent[188].

A new spectral species of indolamineoxygenase having absorption maxima at 415, 542 and 576 nm is observed when potassium superoxide dissolved in dimethylsulfoxide is slowly infused into a solution containing the native ferric form of enzyme[189].

Available evidence indicates that the new spectral species is the catalytically active ferric heme-superoxide $[Fe(III)O_2^-]$ complex that, in turn, is in resonance equilibrium with the ferrous heme-oxygen $[Fe(II)O_2]$ complex[189]. Although the oxygenated form of enzyme is observed in the absence of the organic substrate, steady-state kinetic analyses reveal that the enzyme follows an ordered sequential Bi-Uni mechanism in which the organic substrate is bound to the enzyme before O_2^- [186]. This sequence is analogous to that reported with Pseudomonad tryptophan 2,3-dioxygenase[58] and protocatechuate 3,4-dioxygenase[100].

Brady suggested from the experiments with copper chelators[190] that indoleamine 2,3-dioxygenase is a copper-containing hemoprotein. On the other hand, Hirata et al. found negligible amounts of copper in their highly purified and active preparation of the enzyme[191].

C) Lipoxygenase from *Fusarium Oxysporum*

Lipoxygenase (EC 1.13.11.12) is an enzyme that catalyzes the hydroperoxidation of polyunsaturated fatty acids and esters containing a *cis-cis*-1,4-pentadiene system (Table 6). In 1947, Theorell et al. obtained the enzyme in a crystalline form from soybeans and reported that the enzyme neither contained nor required a metal cofactor[192]. Subsequent studies from three groups of investigators have demonstrated that the enzyme purified from soybeans in an iron-containing dioxygenase[74-76].

Recently, however, a lipoxygenase-like enzyme has been found to be produced by a fungus, identified as *Fusarium oxysporum*[162]. The enzyme has been purified to homogeneity from the extracts of the fungus[163]. The molecular weight of the enzyme is estimated between 12,000 and 13,000 on the basis of ultracentrifugation, SDS-polyacrylamide gel electrophoresis and gel filtration. The purified enzyme shows the typical absorption spectra of a hemoprotein with an absorption maxima at 413 nm and two broad peaks at 530 and 560 nm in the ferric state and maxima at 423 nm, at 525 nm with shoulders around 530 nm, 552 nm and 561 nm in the ferrous state. Identification and quantitative determination of the heme reveals that the purified enzyme possesses 1 mole of protoheme IX per mole of enzyme as a prosthetic group[163]. The enzyme requires Co^{2+} as a stabilizing factor. The isoelectric point of the enzyme is 9.46, and the optimum pH of the reaction is about 12. The enzyme catalyzes the peroxidation of the following compounds by the relative rate shown in parenthesis, linoleic acid (100), methyllinoleate (49), linolenic acid (10), and methyllinolenate (5). Various chelators including o-phenanthioline, 8-hydroxyquinoline, thioglycollic acid, and quinalizarin inhibit the enzyme activity. Other substances such as azide, fluoride and p-chloromercuribenzoate are ineffective, whereas cyanide is a highly effective inhibitor[163].

These results indicate that the *Fusarium* lipoxygenase differs from the soybean lipoxygenase in various respects; soybean lipoxygenase is a nonheme iron-containing dioxygenase and has a molecular weight of 102,000, optimum pH of 6.5 to 7.0 and isoelectric point of pH 5.4. The soybean enzyme is not inhibited by cyanide and catalyzes the peroxidation of linoleic acid and linolenic acid at equal rates[74-76, 193].

V. Copper-Containing Dioxygenase

Quercetinase (Quercetin 2,3-dioxygenase. EC 1.13.11.24) is the only dioxygenase known to contain copper as the sole cofactor[194]. The enzyme catalyzes the reaction shown in Eq. (21).

$$(21)$$

Quercetin (*3*) is converted to depside (*4*) and carbon monoxide with incorporation of two atoms of molecular oxygen. A tracer experiment indicates that an oxygen molecule is incorporated into *4*, but not into carbon monoxide[195, 196].

Quercetinase is an induced extracellular dioxygenase produced by *Aspergillus flavus* when grown on rutin, a flavonoid glycoside[197]. The dioxygenase has been purified from the culture media to a homogeneous protein[198]. The molecular weight of the enzyme is determined as 111,000 ± 4,000. The enzyme is a glycoprotein containing 27.5% carbohydrate[198], and is colorless except in highly concentrated solutions containing 30% protein. Such solutions have a very pale greenish color. The absorption spectrum has a single peak with a maximum at 280 nm and no significant absorption is observed with a sample containing 26 mg of protein per ml between 350 and 800 nm[194]. However, atomic absorption measurements reveal that copper is the only metal present in the enzyme in significant and consistent concentrations, and is equal to about 2 g-atoms per mole of enzyme. The copper appears to be in the cupric form. Compounds such as ethylxanthate, diphenylthiocarbazone, toluene-3,4-dithiol and diethyldithiocarbamate that are highly specific chelators for copper inhibit the dioxygenase at low concentrations, whereas iron chelators are ineffective. These results indicate that quercetinase is a copper-containing dioxygenase and that the copper is related to the enzyme activity[194].

VI. Flavin-Containing Dioxygenases

A) 2-Methyl-3-Hydroxypyridine 5-Carboxylic Acid Dioxygenase

The enzyme was isolated from a *Pseudomonas sp. MA-1* and crystallized by Sparrow *et al.*[61]. The enzyme catalyzes the conversion of 2-methyl-3-hydroxypyridine-5-car-boxylic acid (*5*) to α-(N-acetylaminomethylene)-succinic acid (*6*) [Eq. (22)].

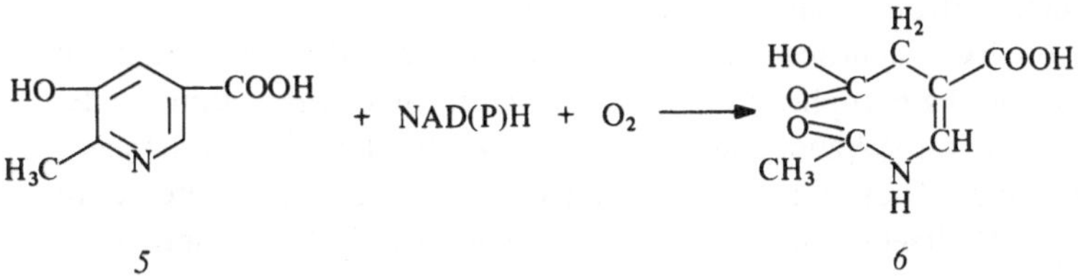

$$\tag{22}$$

The crystalline oxygenase is homogeneous, and contains 2 moles of FAD per mole of enzyme, based on the molecular weight of 166,000. The activity is lost on resolution with acidic ammonium sulfate and can be completely restored with FAD, but not with FMN. Under anaerobic conditions, the FAD in the enzyme is reduced by NADH, $NaBH_4$ or $Na_2S_2O_4$; admission of oxygen reoxidizes the enzyme slowly, whereas on addition of oxygen and compound *5*, rapid reoxidation of the enzyme occurs and the open chain product appears[61]. A tracer experiment indicates that the enzyme is classified as a dioxygenase, and the overall reaction shown in Eq.(22) appears to be catalyzed by the dioxygenase in a concerted fashion *via* a ternary complex of the enzyme, NADH and compound *5*. NADPH is almost as effective as NADH in the reaction[61].

B) 2-Nitropropane Dioxygenase

The finding of 2-nitropropane dioxygenase, an enzyme that catalyzes the following reaction [Equation (23)], has recently been reported by Kido *et al.*[199].

$$
\begin{array}{ccc}
\text{CH}_3 & & \text{CH}_3 \\
| & & | \\
2 \text{ HC--NO}_2 + \text{O}_2 & \rightarrow & 2 \text{ C=O} + 2 \text{ HNO}_2 \\
| & & | \\
\text{CH}_3 & & \text{CH}_3
\end{array}
\tag{23}
$$

Stoichiometrical studies together with the results of ^{18}O experiments show that 2 atoms of molecular oxygen are incorporated into two molecules of acetone formed from 2-nitropropane. Accordingly, the enzyme is an intermolecular dioxygenase and is unique in incorporating 2 atoms of the oxygen molecule into 2 molecules of the same acceptor[199].

The enzyme purified to homogeneity from *Hansenula mrakii* (IFO 0895) has a molecular weight of approximately 62,000 and consists of two subunits non-identical in molecular weight (39,000 and 25,000). The enzyme exhibits absorption

maxima at 274, 370, 415 and 440 nm and a shoulder at 470 nm, and contains
1 mole of FAD and 1 g atom of nonheme iron per mole of enzyme. The enzyme-
bound FAD is reduced by the substrate, 2-nitropropane, under anaerobic conditions,
but the enzyme-bound Fe(III) is not affected. The enzyme can catalyze the oxygenation of the following nitroalkane compounds with the relative rates shown in parenthesis; 2-nitropropane (100), nitroethane (88.7), 3-nitro-2-pentanol (40.6), 1-nitro-
propane (23.4), 3-nitro-2-butanol (13.8), 3-nitropropionate (11.7), and 2-nitro-1-
butanol (2.7). These substrates are converted into nitrite and the corresponding
carbonyl compounds. Nitromethane is not oxygenated at all[199].

The enzyme activity is significantly inhibited by Tiron and 8-hydroxyquinoline,
but not by α,α'-dipyridyl and o-phenanthroline. The addition of thiol compounds
such as cysteine, 2-mercaptoethanol and glutathione, and thiol inhibitors such as
p-chloromercuribenzoate, N-ethylmaleimide and $HgCl_2$ also markedly decreases the
enzyme activity. The Michaelis constants of the enzyme are as follows: 2-nitropro-
pane ($2.13 \times 10^{-2} M$), nitroethane ($2.43 \times 10^{-2} M$), 3-nitro-2-pentanol ($6.8 \times 10^{-3} M$),
1-nitropropane (2.56×10^{-2} M) and oxygen ($3.63 \times 10^{-4} M$ with 2-nitropropane)[199].

VII. Dioxygenases Involved in the Formation of Catechol Derivatives

Catechol and its derivatives are known to be intermediates in the microbial degradation of various aromatic compounds. Several enzymes or enzyme systems are reported
to catalyze the formation of catechol or its derivatives from various aromatic compounds including anthranilate, benzoate, benzene and pyrazone. These reactions are
initiated by the oxygenation of their phenol moiety to the corresponding *cis,ortho*-
dihydrodihydroxy compounds which are then converted to the corresponding catechol derivatives by dehydrogenation[200]. These initial oxygenation reactions are
catalyzed by individual dioxygenases. These are listed in Table 7.

The dioxygenase nature of these reactions was first demonstrated by Kobayashi
et al.[201] with anthranilate 1,2-dioxygenase (hydroxylase). They demonstrated by
experiments with [18]O that both atoms of oxygen in catechol are exclusively derived
from molecular oxygen. Subsequently, the incorporation of two atoms of molecular
oxygen into substrates was established by tracer experiments with benzene 1,2-dioxy-
genase[204] and benzoate 1,2-dioxygenase[206].

A) Benzene 1,2-Dioxygenase

Benzene 1,2-dioxygenase was first demonstrated in cell-free extracts of a strain of
Pseudomonas putida by Gibson *et al.*[203]. They were able to resolve the system into
two fractions by $(NH_4)_2SO_4$ precipitation, both of which are necessary for benzene
oxidation. Subsequently, Axcell and Geary[82] obtained a soluble enzyme system
which oxidizes benzene to *cis*-1,2-dihydrocyclohexa-3,5-diene (*cis*-benzeneglycol)
from a species of *Pseudomonas* grown on benzene as the major carbon source. The
system is shown to consist of three protein components. Two of these are nonheme

Table 7. Dioxygenases involved in the formation of catechol derivatives

Enzymes	EC Number		Cofactor	Ref.
Anthranilate 1,2-dioxygenase (deaminating, decarboxylating)	1.14.12.1		Fe(II), NADH	201)
Anthranilate 2,3-dioxygenase (deaminating)	1.14.12.2		Fe(II), NADPH	202)
Benzene 1,2-dioxygenase	1.14.12.3		Fe(II), NADH	82, 203), 204)
Benzoate 1,2-dioxygenase	1.13.99.2		Fe(II), NADH	62, 205, 206)
Pyrazon dioxygenase			Fe(II), NADH	83)

175

iron proteins of molecular weight approx. 21,000 and 186,000, and the other is a flavoprotein of molecular weight approx. 60,000. Fe(II) and NADH are essential cofactors for the benzene oxidation[82]. They demonstrated that the 186,000 dalton protein (Al) is the terminal dioxygenase, and suggested the functions of the components as shown in Eq. (24).

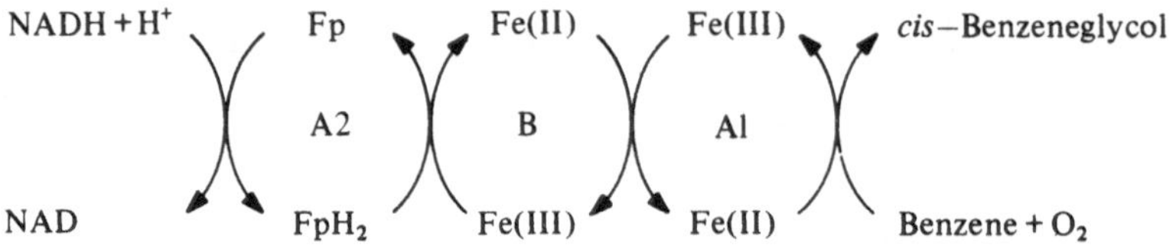

$$\text{NADH} + \text{H}^+ \quad \text{Fp} \qquad \text{Fe(II)} \qquad \text{Fe(III)} \qquad cis-\text{Benzeneglycol}$$

A2   B   Al   (24)

$$\text{NAD} \qquad \text{FpH}_2 \qquad \text{Fe(III)} \qquad \text{Fe(II)} \qquad \text{Benzene} + \text{O}_2$$

Where A2 and B are the flavoprotein (Fp) and the small nonheme iron protein, respectively.

B) Benzoate 1,2-Dioxygenase

Benzoate 1,2-dioxygenase was shown to catalyze the conversion of benzoate to 1,2-dihydro-1,2-dihydroxybenzoic acid (DHB) in the presence of NADH and oxygen by Reiner et al.[205, 206]. Yamaguchi et al.[62] have demonstrated that the enzyme system consists of two components (A and B), both of which are required for benzoate 1,2-dioxygenase activity. Component A shows NADH-cytochrome c reductase activity and component B appears to be dioxygenase, thus the following reaction scheme is suggested [Eq. (25)].

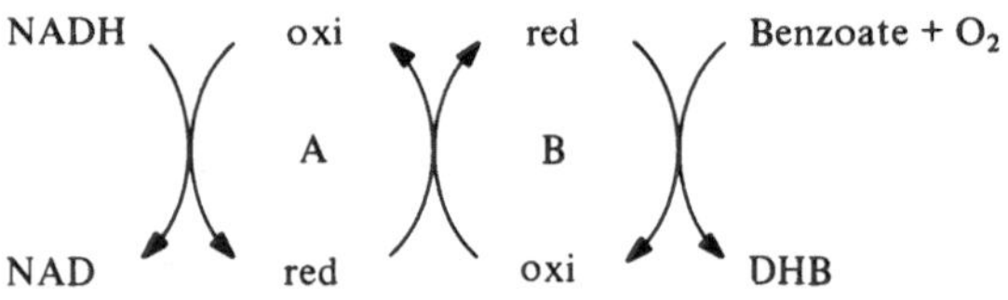

$$\text{NADH} \qquad \text{oxi} \qquad \text{red} \qquad \text{Benzoate} + \text{O}_2$$

A   B   (25)

$$\text{NAD} \qquad \text{red} \qquad \text{oxi} \qquad \text{DHB}$$

C) Pyrazon Dioxygenase

Pyrazon dioxygenase was also shown to consist of three different enzyme components[83]. No component alone oxidizes the phenol moiety of pyrazon; only when the three components are combined can oxidation be detected. Component Al is an iron-sulfur protein (MW about 180,000, red-brown in color). Two g-atoms of iron and two moles of inorganic sulfur are present per 180,000 g of the protein. Component A2 is a yellow protein of a molecular weight of about 67,000. FAD was shown to be the prosthetic group of this protein. Component B (MW about 12,000, brown in color) is a protein of the ferredoxin type. In accordance with the benzene-oxidizing system[82] and with a monooxygenase system[48, 49], Sanber et al. propose the hypothetical scheme shown in Eq. (26)[83].

$$\text{NADH} + \text{H}^+ \xrightarrow{\text{Fp}} \xrightarrow{\text{A2}} \xrightarrow{\text{Fe(II)}} \xrightarrow{\text{B}} \xrightarrow{\text{Fe(III)}} \xrightarrow{\text{Al}} \text{cis-Pyrazon dihydrodiol}$$

$$\text{NAD} \quad \text{FpH}_2 \quad \text{Fe(III)} \quad \text{Fe(II)} \quad \text{Pyrazon} + \text{O}_2 \tag{26}$$

VIII. α-Ketoglutarate-Requiring Dioxygenases

The reactions catalyzed by the α-ketoglutarate-requiring dioxygenases are listed in Table 8. These reactions are known to require α-ketoglutarate in addition to ferrous iron and ascorbic acid. The requirement of the ferrous iron is highly specific, whereas ascorbic acid can be replaced by other reductants including some sulfhydryl containing compounds. Subsequent studies on the role of α-ketoglutarate revealed that the compound acts as a cosubstrate and is converted to succinate with incorporation of one atom of molecular oxygen and decarboxylation. The overall reaction may be schematically shown by Eq. (11). For the details concerning these enzymes, the readers are referred to the recent review articles[14, 219].

The mechanism of α-ketoglutarate participation in the dioxygenase reaction has been proposed by Lindstedt and his co-workers[65, 220] as shown in Eq. (27) with γ-butyrobetaine hydroxylase.

$$R_1 = -CH_2-\overset{+}{\underset{CH_3}{\overset{CH_3}{N}}}-CH_3 \quad R_2 = -CH_2-COO^- \quad R_3 = -CH_2-CH_2-COO^-$$

In this mechanism, the peroxide anion of the substrate makes a nucleophilic attack on the carbonyl carbon of α-ketoglutarate so that a peroxide bridge is formed between the two compounds. In this way, one atom each of molecular oxygen is incorporated into the substrate and α-ketoglutarate to form the product and succinate. The incorporation of molecular oxygen into succinate has been shown in most of the reactions that require α-ketoglutarate.

177

Table 8. α-Ketoglutarate requiring dioxygenases[a]

Enzyme	EC Number	Reaction	Cofactor	Ref.
Proline, 2 oxoglutarate dioxygenase (Prolyl hydroxylase)	1.14.11.2	$+ \alpha KG + O_2 \longrightarrow + Succ. + CO_2$	Fe(II), Asc.	207–210
Lysine, 2-oxoglutarate dioxygenase (Lysyl hydroxylase)	1.14.11.4	$\begin{array}{c} CH_2NH_2 \\ CH_2 \\ (CH_2)_2 \\ R-NH-CH-COR' \end{array} + \alpha KG + O_2 \longrightarrow \begin{array}{c} CH_2-NH_2 \\ CHOH \\ (CH_2)_2 \\ R-NH-CH-COR' \end{array} + Succ. + CO_2$	Fe(II), Asc.	211, 212
γ-Butyrobetaine, 2 oxoglutarate dioxygenase (γ-Butyrobetaine hydroxylase)	1.14.11.1	$(CH_3)_3\overset{+}{N}CH_2-CH_2-CH_2-COOH + \alpha KG + O_2 \longrightarrow (CH_3)_3\overset{+}{N}CH_2CHOHCH_2COOH + Succ. + CO_2$	Fe(II), Asc.	63–65
Thymidine, 2 oxoglutarate dioxygenase (Pyrimidine deoxyribonucleoside 2'-hydroxylase)	1.14.11.3	$+ \alpha KG + O_2 \longrightarrow + Succ. + CO_2$	Fe(II), Asc.	213, 214

Thymine, 2 oxoglutarate dioxygenase
(Thymine-7-hydroxylase)   1.14.11.6

$+ \alpha KG + O_2 \longrightarrow$ $+ \text{Succ.} + CO_2$ Fe(II), Asc.   215, 216)

5-Hydroxymethyluracil 2-oxo-
glutarate dioxygenase
(5-hydroxymethyluracil oxygenase)   1.14.11.5

$+ \alpha KG + O_2 \longrightarrow$ $+ \text{Succ.} + CO_2$ Fe(II), Asc.   217)

4-Hydroxyphenylpyruvate
dioxygenase   1.13.11.27

$+ O_2 \longrightarrow$ $+ CO_2$ Fe(II), Asc.   218)

a   The following abbreviations are used; α-KG: α-ketoglutarate; Succ.: succinate; Asc.: Ascorbic acid.

The reaction catalyzed by p-hydroxyphenylpyruvate dioxygenase does not require α-ketoglutarate. However, the substrate, p-hydroxyphenylpyruvate, has an internal keto acid in the molecule. The tracer experiments reveal that one atom of molecular oxygen is incorporated into the carbonyl group with simultaneous decarboxylation, and the other atom of oxygen into the newly formed hydroxyl group of the product, homogentisate[218]. Thus, the reaction mechanism of the enzyme is quite similar to those of α-ketoglutarate-requiring enzymes. A reaction mechanism for the enzyme in which a cyclic peroxide (7) and a quinol (8) are involved as intermediates was proposed by Witkop and his co-workers[221, 222] [Mechanism (a) in Eq. (28)] and another mechanism [(b) in Eq. (28)] involving a peracid (9) as an intermediate was also proposed by Hamilton[223].

Mechanism (a)

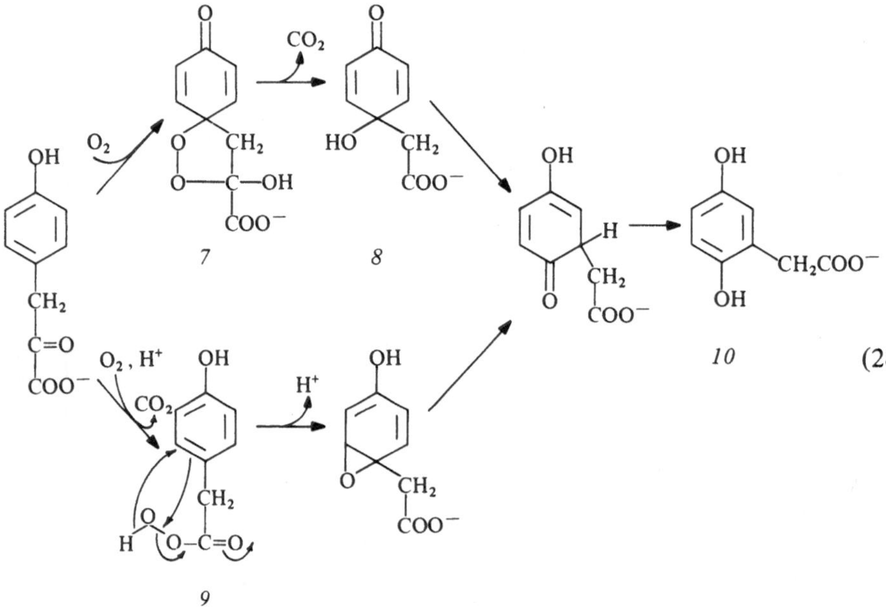

$$(28)$$

Mechanism (b)

On the other hand, the chemically synthesized quinol intermediate (8) is shown to be converted nonenzymatically to homogentisate (10) in an alkaline solution by Saito et al.[224]. However, the compound (8) cannot be demonstrated to be an enzymatically active intermediate of the reaction[225, 226].

IX. Concluding Remarks

Dioxygenases have been discovered in all types of living organisms and shown to catalyze a variety of reactions, including ring cleavage reactions, hydroxylation reactions, oxidation of certain sulfur containing compounds and so forth. Among these, the cleavage of the aromatic ring appears to depend largely or entirely upon this type of enzyme. Thus, extensive studies on the reaction mechanism of dioxygenase have been carried out, especially with microbial dioxygenases that catalyze the cleavage of the aromatic rings.

Iron appears to be the most common cofactor of dioxygenases and is in the form of ferrous, ferric and heme iron. Some enzymes contain copper as a cofactor and some flavin. The flavin-containing dioxygenases may also contain iron as the second cofactor. These metal cofactors appear to be a site of oxygen activation as well as closely related to the substrate binding site. An oxygenated form of enzyme has been demonstrated to be an obligatory intermediate with several iron-containing dioxygenases. In all cases, the enzyme reacts with the organic substrate first and then reacts with oxygen to form a ternary complex of enzyme, substrate and oxygen, as an oxygenated intermediate. Both substrate and oxygen are presumed to be activated in the ternary complex and react together to form the oxygenated end product. Although the activation mechanism remains to be elucidated, the structural analyses of the enzyme as well as the enzymatically active species of the oxygenated intermediate may open the way to understanding what the active form of oxygen is in the biological system. These analyses by means of not only chemical methods but also physical methods such as ESR, Mössbauer, NMR and Raman spectroscopies, X-ray crystallography, and so forth are now being carried out in several laboratories including our own. Thus, thorough understanding of the reaction mechanism of these enzymes can only be achieved by intimate collaborations of scientists in various fields including biochemists, organic chemists, inorganic chemists and physicists.

Acknowledgements. The author would like to express his sincere thanks to Dr. Yukikazu Saeki for his critical reading of this manuscript. Thanks are also due to Mr. Peter Schneider for his help in the preparation of the manuscript. In addition, the supports in part by a grant from the Naito Foundation and by a Grant-in-aid for Scientific Research from the Ministry of Education, Science and Culture, Japan are gratefully acknowledged.

X. References

1. Hayaishi, O., in Oxygenases. Hayaishi, O., (ed.). New York: Academic Press, 1962, p. 1
2. Mason, H. S., Fowlks, W. L., Peterson, L.: J. Amer. Chem. Soc. 77, 2914 (1955)
3. Hayaishi, O., Katagiri, M., Rothberg, S.: J. Amer. Chem. Soc. 77, 5450 (1955)
4. Hayaishi, O., Rothberg, S., Methler, A. H.: Abstr. 130th Meeting Amer. Chem. Sog., p 53C (1956)
5. Hayaishi, O. (ed.): Oxygenases. New York: Academic Press, 1962
6. Bloch, K., Hayaishi, O. (eds.): Biological and chemical aspects of oxygenases. Tokyo: Maruzen, 1966
7. Hayaishi, O. (ed.): Molecular mechanisms of oxygen activation. New York: Academic Press, 1974
8. Mason, H.S.: Adv. Enzymol. *19*, 79 (1957)
9. Hayaishi, O.: Ann. Rev. Biochem. *31*, 25 (1962)
10. Mason, H. S.: Ann. Rev. Biochem. *34*, 595 (1965)

11. Hayaishi, O.: Pharmacol. Rev. *18*, 71 (1966)
12. Hayaishi, O., Nozaki, M.: Science *164*, 389 (1969)
13. Nozaki, M., Ishimura, Y., in Microbial iron metabolism. Nielands, J. B. (ed.). New York: Academic Press, 1975, p. 417
14. Hayaishi, O., Nazoki, M., Abbott, M. T.: The enzymes Boyer, P. (ed.). New York: Academic Press, 1975, Vol. XII, p. 119
15. Hayaishı, O.: Proc. Plenary Sessions, 6th Internatl. Congr. Biochem. New York, Vol. 33, p. 31 (1964)
16. Takeda, H., Hayaishi, O.: J. Biol. Chem. *241*, 2733 (1966)
17. Takeda, H., Yamamoto, S., Kojima, Y., Hayaishi, O.: J. Biol. Chem. *244*, 2935 (1969)
18. Yamamoto, S., Takeda, H., Maki, Y., Hayaishi, O.: J. Biol. Chem. *244*, 2951 (1969)
19. Katagiri, M., Yamamoto, S., Hayaishi, O.: J. Biol. Chem. *237*, PC2413 (1962)
20. Yamamoto, S., Katagiri, M., Maeno, H., Hayaishi, O.: J. Biol. Chem. *240*, 3408 (1965)
21. Friedman, S., Kaufman, S.: J. Biol. Chem. *240*, 552 (1965)
22. Friedman, S., Kaufman, S.: J. Biol. Chem. *240*, 4763 (1965)
23. Goldstein, M., Lauber, E., McKereghan, M. R.: J. Biol. Chem. *240*, 2066 (1965)
24. Kaufman, S.: J. Bıol. Chem. *226*, 511 (1957)
25. Kaufman, S.: J. Bıol. Chem. *237*, PC2712 (1962)
26. Kaufman, S.: J. Bıol. Chem. *239*, 332 (1964)
27. Fisher, D. B., Kirkwood, R., Kaufman, S.: J. Biol. Chem. *247*, 5161 (1972)
28. Omura, T., Sato, R.: Biochim. Biophys. Acta *71*, 224 (1963)
29. Omura, T., Sato, R.: J. Biol. Chem. *239*, 2370 (1964)
30. Orrenius, S., Pallner, G., Ernster, L.: Biochem. Biophys. Res. Commun. *14*, 329 (1964)
31. Copper, D. Y., Levın, S., Narasımhulu, S., Rosenthal, O., Estabrook, R. W.: Science *147*, 400 (1965)
32. Lu, A. Y. H., Coon, M. J.: J. Biol. Chem. *243*, 1331 (1968)
33. Oshıno, N., Imai, Y., Sato, R.: Bıochim. Biophys. Acta *128*,13 (1966)
34. Oshıno, N., Imaı, Y., Sato, R.: J. Biochem. *69*, 155 (1971)
35. Strittmatter, P., Spatz, L., Corocoran, D., Rogers, M. J., Setlow, B., Redlıne, R.: Proc. Nat. Acad. Scı. U.S.A. *71*, 4565 (1974)
36. Estabrook, R. W., Cooper, D. T., Rosenthal, O.: Bıochem. Z. *338*, 741 (1963)
37. Suzuki, K., Kimura, T.: Bıochem. Bıophys. Res. Commun. *19*, 340 (1965)
38. Omura, T., Sato, R., Cooper, D. Y., Rosenthal, O., Estabrook, R. W.: Fed. Proc. *24*, 1181 (1965)
39. Omura, T., Saunders, E., Estabrook, R. W., Cooper, D. Y., Rosenthal, O.: Arch. Biochem. Biophys. *117*, 660 (1966)
40. Katagıri, M., Gangulı, B. N., Gunsalus, I. C.: J. Biol. Chem. *243*, 3543 (1968)
41. Tsibrıs, J. C. M., Tsaı, R. L., Gunsalus, I. C., Orme-Johnson, W. H., Hansen, R. E., Beinert, H.: Proc. Nat. Acad. Scı. U.S.A. *59*, 959 (1968)
42. Gunsalus, I. C., Meeks, J. R., Lipscomb, J. D., Debrunner, P. G., Munck, E., ın Molecular mechanısms of oxygen actıvatıon. Hayaishı, O. (ed.)., New York: Academic Press 1974, p.559
43. Peterson, J. A., Kusunose, M., Kusunose, E., Coon, M. J.: J. Biol. Chem. *242*, 4334 (1967)
44. Peterson, J. A., Coon, M. J.: J. Bıol. Chem. *243*, 329 (1968)
45. McKenna, E. J., Coon, M. J.: J. Bıol. Chem. *245*, 3882 (1970)
46. Bernhardt, F.-H., Erdin, N., Staudınger, H., Ullrich, V.: Europ. J. Bıochem. *35*, 126 (1973)
47. Bernhardt, F.-H., Pachowsky, H., Staudinger, H.: Eur. J. Biochem. *57*, 241 (1975)
48. Ziegler, D. M., Mıtchell, C. H.: Arch. Biochem. Biophys. *150*, 116 (1972)
49. Ono, T., Bloch, K.: J. Bıol. Chem. *250*, 1571 (1975)
50. Brook, D. F., Large, P. J.: Europ. J. Bıochem. *55*, 601 (1975)
51. Lennarz, W. J., Bloch, K.: J. Biol. Chem. *235*, PC26 (1960)
52. Lıght, R. T., Lennarz, M. T., Bloch, K.: J. Bıol. Chem. *237*,1793 (1962)
53. Nozakı, M., Kagamiyama, H., Hayaıshı, O.: Biochem. Z. *338*, 582 (1963)
54. Nozakı, M., Ono, K., Nakazawa, T., Kotani, S., Hayaishi, O.: J. Bıol. Chem. *243*, 2682 (1968)
55. Kojıma, Y., Fujısawa, H., Nakazawa, A., Nakazawa, T., Kanetsuna, F., Tanıuchi, H., Nozakı, M., Hayaıshı, O.: J. Biol. Chem. *242*, 3270 (1967)

56. Nakazawa, T., Nozaki, M., Hayaishi, O., Yamano, T.: J. Biol. Chem. *244*, 119 (1969)
57. Tanaka, T., Knox, W. E.: J. Biol. Chem. *234*, 1162 (1959)
58. Ishimura, Y., Nozaki, M., Hayaishi, O., Nakamura, T., Tamura, M., Yamazaki, I.: J. Biol. Chem. *245*, 3593 (1970)
59. Krishnamurty, H. G., Simpson, F. J.: J. Biol. Chem. *245*, 1467 (1969)
60. Oka, T., Simpson, F. J.: Biochem. Biophys. Res. Commun. *43*, 1 (1971)
61. Sparrow, L. G., Ho, P. P. K., Sundaran, T. K., Zach, D., Nyns, E. J., Snell, E. E.: J. Biol. Chem. *244*, 2590 (1969)
62. Yamaguchi, M., Yamauchi, T., Fujisawa, H.: Biochem. Biophys. Res. Commun. *67*, 264 (1975)
63. Lindstedt, G., Lindstedt, S., Olander, B., Tofft, M.: Biochim. Biophys. Acta *158*, 503 (1968)
64. Lindblad, B., Lindstedt, G., Lindstedt, S., Tofft, M.: J. Amer. Chem. Soc. *91*, 4604 (1969)
65. Lindstedt, G., Lindstedt, S.: J. Biol. Chem. *245*, 4187 (1970)
66. Suzuki, I.: Biochim. Biophys. Acta *104*, 359 (1965)
67. Suzuki, I.: Biochim. Biophys. Acta *110*, 97 (1965)
68. Cavallini, D., Scandurra, R., De Marco, C.: J. Biol. Chem. *238*, 2999 (1963)
69. Cavallini, D., Scandurra, R., Monacelli, F.: Biochem. Biophys. Res. Commun. *24*, 185 (1966)
70. Rotilio, G., Federici, G., Calabrese, L., Costa, M., Cavallini, D.: J. Biol. Chem. *245*, 6235 (1970)
71. Sörbo, Bo, Ewetz, L.: Biochem. Biophys. Res. Commun. *18*, 359 (1965)
72. Lombardini, J. B., Turini, P., Biggo, D. R., Singer, T. P.: Physiol. Chem. & Physico. *1*, 1 (1969)
73. Sakakibara, S., Yamaguchi, K., Hosokawa, Y., Kohashi, N., Ueda, I., Sakamoto, Y.: Biochim. Biophys. Acta *422*, 273 (1976)
74. Roza, M., Francke, A.: Biochim. Biophys. Acta *327*, 24 (1973)
75. Chan, H. W. S.: Biochim. Biophys. Acta *327*, 32 (1973)
76. Pistorius, E. K., Axelrod, B.: J. Biol. Chem. *249*, 3183 (1974)
77. Nugteren, D. H., Beerthuis, R. K., van Dorp, D. A.: Rec. Trav. Chim. *85*, 405 (1966)
78. Samuelsson, B.: Prog. Biochem. Pharmacol. *3*, 59 (1967)
79. Yoshimoto, A., Ito, H., Tomita, K.: J. Biochem. *68*, 487 (1970)
80. Miyamoto, T., Ogino, N., Yamamoto, S., Hayaishi, O.: J. Biol. Chem. *251*, 2629 (1976)
81. Nozaki, M., in: Molecular mechanisms of oxygen activation. Hayaishi, O. (ed.). New York: Academic Press, 1974, p. 135
82. Axcell, B. C., Geary, P. J.: Biochem. J. *146*, 173 (1975)
83. Sanber, K., Frökner, C., Rosenberg, G., Eberspächer, J., Lingens, F.: Eur. J. Biochem. *74*, 89 (1977)
84. Abbott, M. T., Udenfriend, S., in: Molecular mechanisms of oxygen activation. Hayaishi, O. (ed.). New York: Academic Press, 1974, p. 168
85. Nozaki, M., Kotani, S., Ono, K., Senoh, S.: Biochim. Biophys. Acta *220*, 213 (1970)
86. Stanier, R. Y., Ingraham, J. L.: J. Biol. Chem. *210*, 799 (1954)
87. Dagley, S., Stopher, D. A.: Biochem. J. *73*, 16P (1959)
88. Kojima, Y., Itada, N., Hayaishi, O.: J. Biol. Chem. *236*, 2223 (1961)
89. Adachi, K., Takeda, Y., Senoh, S., Kita, H.: Biochim. Biophys. Acta *93*, 483 (1964)
90. Kita, H.: J. Biochem. *58*, 116 (1965)
91. Ono-Kamimoto, M.: J. Biochem. *74*, 1049 (1973)
92. Dagley, S., Patel, M. D.: Biochem. J. *66*, 227 (1957)
93. Cain, R. B.: Nature *193*, 842 (1962)
94. Ono, K., Nozaki, M., Hayaishi, O.: Biochim. Biophys. Acta *220*, 224 (1970)
95. Nakazawa, A., Nakazawa, T., Kotani, S., Nozaki, M., Hayaishi, O.: J. Biol. Chem. *244*, 1527 (1969)
96. Fujiwara, M., Golovleva, L. A., Saeki, Y., Nozaki, M., Havaishi, O.: J. Biol. Chem. 250, 4848 (1975)
97. Nozaki, M., Yoshida, R., Nakai, C., Iwaki, M., Saeki, Y., Kagamiyama, H., in: Iron and copper proteins. Yasunobu, K. T., Mower, H. F., Hayaishi, O. (eds.) Adv. Exp. Med. Biol. Vol. 74. New York: Plenum Press, 1976, p. 127

98. Fujisawa, H., Hayaishi, O.: J. Biol. Chem. *243*, 2673 (1968)
99. Fujisawa, H., Uyeda, M., Kojima, Y., Nozaki, M., Hayaishi, O.: J. Biol. Chem. *247*, 4414 (1972)
100. Fujisawa, H., Hiromi, K., Uyeda, M., Okuno, S., Nozaki, M., Hayaishi, O.: J. Biol. Chem. *247*, 4422 (1972)
101. Fujiwara, M., Nozaki, M.: Biochim. Biophys. Acta *327*, 306 (1973)
102. Zaborsky, O. R., Hou, C. T., Ogletree, J.: Biochim. Biophys. Acta *386*, 18 (1975)
103. Hou, C. T.: Biochemistry *14*, 3899 (1975)
104. Que, Jr., L., Lipscomb, J. D., Zimmermann, R., Münck, E., Orme-Johnson, N. R., Orme-Johnson, W. H.: Biochim. Biophys. Acta *452*, 320 (1976)
105. Yoshida, R., Hori, K., Fujiwara, M., Saeki, Y., Kagamiyama, H., Nozaki, M.: Biochemistry *15*, 4048 (1976)
106. Hirata, F., Nakazawa, A., Nozaki, M., Hayaishi, O.: J. Biol. Chem. *246*, 5882 (1971)
107. Kita, H., Miyake, Y., Kamimoto, M., Senoh, S., Yamano, T.: J. Biochem. *66*, 45 (1969)
108. Ono-Kamimoto, M., Senoh, S.: J. Biochem. *75*, 321 (1974)
109. Tai, H. H., Sih, C. J.: J. Biol. Chem. *245*, 5062 (1970)
110. Tai, H. H., Sih, C. J.: J. Biol. Chem. *245*, 5072 (1970)
111. Nakai, C., Saeki, Y., Kagamiyama, H., Nozaki, M.: unpublished data
112. Nakagawa, H., Inoue, H., Takeda, Y.: J. Biochem. *54*, 65 (1963)
113. Nagami, K., Miyake, Y.: Biochem. Biophys. Res. Commun. *46*, 198 (1972)
114. Patel, R. N., Hou, C. T., Felix, A., Lillard, M. O.: J. Bacteriol. *127*, 536 (1976)
115. Nakazawa, T., Kojima, Y., Fujisawa, H., Nozaki, M., Hayaishi, O.: J. Biol. Chem. *240*, PC3225 (1965)
116. Nagami, K., Miyake, Y.: Biochem. Biophys. Res. Commun. *42*, 497 (1971)
117. Hou, C. T., Lillard, M. O., Schwartz, R. O.: Biochemistry *15*, 582 (1976)
118. Fujisawa, H., Hiromi, K., Uyeda, M., Nozaki, M., Hayaishi, O.: J. Biol. Chem. *246*, 2320 (1971)
119. Peisach, J., Fujisawa, H., Blumberg, W. H., Hayaishi, O.: Fed. Proc., Fed. Amer. Soc. Exp. Biol. *31*, 448 (1972)
120. Blumberg, W. E., Peisach, J.: Ann. N. Y. Acad. Sci. *222*, 539 (1973)
121. Hou, C. T., Lillard, M. O.: J. Bacteriol. *126*, 516 (1976)
122. Hori, K., Hashimoto, T., Nozaki, M.: J. Biochem. *74*, 375 (1973)
123. Nozaki, M., Nakazawa, T., Fujisawa, H., Kotani, S., Kojima, Y., Hayaishi, O.: Adv. Chem. Ser., No. 77, Oxidation of Organic Compounds III, Amer. Chem. Soc. p. 242, 1968
124. Jamaluddin, M. P.: J. Bacteriol. *129*, 690 (1977)
125. Dagley, S., Geary, P. J., Wood, J. M.: Biochem. J. *109*, 559 (1968)
126. Zabinski, R., Munck, E., Champion, P. M., Wood, J. M.: Biochemistry *11*, 3212 (1972)
127. Mehler, A. H., in: Oxygenases. Hayaishi, O. (ed.). New York: Academic Press, 1962, p. 87
128. Priest, R. E., Bokman, A. H., Schweigert, B. S.: Proc. Soc. Exp. Biol. Med. *78*, 477 (1951)
129. Stevens, C. O., Henderson, L. M.: J. Biol. Chem. *234*, 1188 (1959)
130. Mitchell, E. A., Kang, H. H., Henderson, L. M.: J. Biol. Chem. *238*, 1151 (1963)
131. Ogasawara, N., Gander, J. E., Henderson, L. M.: J. Biol. Chem. *241*, 613 (1966)
132. Koontz, W. A., Shiman, R.: J. Biol. Chem. *251*, 368 (1976)
133. Suda, M., Takeda, Y.: J. Biochem. *37*, 375 (1950)
134. Suda, M., Takeda, Y.: J. Biochem. *37*, 381 (1950)
135. Crandall, D. I.: Fed. Proc. Fed. Amer. Soc. Exp. Biol. *12*, 192 (1953)
136. Schepartz, B.: J. Biol. Chem. *205*, 185 (1953)
137. Knox, W. E., Edwards, S. W.: J. Biol. Chem. *216*, 479 (1955)
138. Crandall, D. I., Kmeger, R. C., Anan, F., Yasunobu, K., Mason, H. S.: J. Biol. Chem. *235*, 3011 (1960)
139. Flamm, W. G., Crandall, D. I.: J. Biol. Chem. *238*, 389 (1963)
140. Adachi, K., Iwayama, Y., Tanioka, H., Takeda, Y.: Biochim. Biophys. Acta *118*, 88 (1966)
141. Takemori, S., Furuya, E., Mihara, K., Katagiri, M.: Eur. J. Biochem. *6*, 411 (1968)
142. Lack, L.: Biochim. Biophys. Acta *34*, 117 (1959)
143. Sugiyama, S., Yano, Y., Arima, K.: Bull. Agr. Chem. Soc. *24*, 243 (1960)

144. Sugiyama, S., Yano, Y., Arima, K.: Bull. Agr. Chem. Soc. *24*, 249 (1960)
145. Crawford, R. L., Hutton, S. W., Chapman, P. J.: J. Bacteriol. *121*, 794 (1975)
146. Dagley, S., Chapman, P. J., Gibson, D. T.: Biochem. J. *97*, 643 (1965)
147. Behrman, E. J., Stanier, R. Y.: J. Biol. Chem. *228*, 923 (1957)
148. Ganthier, J. J., Rittenberg, S. C.: J. Biol. Chem. *246*, 3737 (1971)
149. Ganthier, J. J., Rittenberg, S. C.: J. Biol. Chem. *246*, 3743 (1971)
150. Fujioka, M., Wada, H.: Biochim. Biophys. Acta *158*, 70 (1968)
151. White, G. A., Krupka, R. M.: Arch. Biochem. Biophys. *110*, 448 (1965)
152. Cavallini, D., De Marco, C., Mondovi, B.: Nature *192*, 557 (1961)
153. Cavallini, D., De Marco, C., Scandurra, R.: Ital. J. Biochem. *11*, 196 (1962)
154. Cavallini, D., Scandurra, R., De Marco, C.: Biochem. J. *96*, 781 (1965)
155. Cavallini, D., De Marco, C., Scandurra, R., Dupré, S., Graziani, M. T.: J. Biol. Chem. *241*, 3189 (1966)
156. Cavallini, D., Cannella, C., Federici, G., Dupré, S., Fiori, A., Grosso, E. D.: Eur. J. Biochem. *16*, 537 (1970)
157. Federici, G., Barra, D., Fiori, A., Costa, M.: Physiol. Chem. Phys. *3*, 448 (1971)
158. Wainer, A.: Biochem. Biophys. Acta *128*, 296 (1966)
159. Lombardini, J. B., Singer, T. P., Boyer, P. D.: J. Biol. Chem. *244*, 1172 (1969)
160. Hou, L. T., Patel, R., Lillard, M. O.: Applied Environmental Microbiol. *33*, 725 (1977)
161. Feigelson, P., Brady, F. O., in: Molecular mechanism of oxygen activation. Hayaishi, O. (ed.). New York: Academic Press, 1974, p. 87.
162. Satoh, T., Matsuda, Y., Takashio, M., Satoh, K., Beppu, T., Arima, K.: Agr. Biol. Chem. *40*, 953 (1976)
163. Matsuda, Y., Satoh, T., Beppu, T., Arima, K.: Agr. Biol. Chem. *40*, 963 (1976)
164. Kotake, Y., Masayama, T.: Hoppe-Seyler's Z. Physiol. Chem. *243*, 237 (1936)
165. Knox, W. E., Mehler, A. H.: J. Biol. Chem. *187*, 419 (1950)
166. Hayaishi, O., Stanier, R. Y.: J. Bacteriol. *62*, 691 (1951)
167. Poillon, W. N., Maeno, H., Koike, K., Feigelson, P.: J. Biol. Chem. *244*, 3447 (1969)
168. Schutz, G., Feigelson, P.: J. Biol. Chem. *247*, 5327 (1972)
169. Brady, F. O., Monaco, M. E., Forman, H. J., Schutz, G., Feigelson, P.: J. Biol. Chem. *247*, 7915 (1972)
170. Maeno, H., Feigelson, P.: Biochem. Biophys. Res. Commun. *21*, 297 (1965)
171. Ishimura, Y., Hayaishi, O.: J. Biol. Chem. *248*, 8610 (1973)
172. Forman, H. J., Feigelson, P.: Biochemistry *10*, 760 (1971)
173. Brady, F. O., Forman, H. J., Feigelson, P.: J. Biol. Chem. *246*, 7119 (1971)
174. Brady, F. O., Feigelson, P., Rajagopalan, K. V.: Arch. Biochem. Biophys. *157*, 63 (1973)
175. Brady, F. O., Feigelson, P.: J. Biol. Chem. *250*, 5041 (1975)
176. Makino, R., Ishimura, Y.: J. Biol. Chem. *251*, 7722 (1976)
177. Ishimura, Y., Hayaishi, O., Peisach, J., Blumberg, W. E.: (1974) in preparation (see also Ref. [131])
178. Ishimura, Y., Nozaki, M., Hayaishi, O., Tamura, M., Yamazaki, I.: J. Biol. Chem. *242*, 2574 (1967)
179. Ishimura, Y., Nozaki, M., Hayaishi, O., Tamura, M., Yamazaki, I.: Advan. Chem. *77*, 235 (1968)
180. Kotake, Y., Ito, N.: J. Biochem. *25*, 71 (1937)
181. Higuchi, K., Kuno, S., Hayaishi, O.: Arch. Biochem. Biophys. *120*, 397 (1967)
182. Yamamoto, S., Hayaishi, O.: J. Biol. Chem. *242*, 5260 (1967)
183. Hirata, F., Hayaishi, O.: Biochem. Biophys. Res. Commun. *47*, 1112 (1972)
184. Hirata, F., Hayaishi, O., Tokuyama, T., Senoh, S.: J. Biol. Chem. *249*, 1311 (1974)
185. Hirata, F., Hayaishi, O.: J. Biol. Chem. *246*, 7825 (1971)
186. Ohnishi, T., Hirata, F., Hayaishi, O.: J. Biol. Chem. *252*, 4643 (1977)
187. Nishikimi, M.: Biochem. Biophys. Res. Commun. *63*, 92 (1975)
188. Hayaishi, O., Hirata, F., Ohnishi, T., Henry, J.-P., Rosenthal, I., Katoh, A.: J. Biol. Chem. *252*, 3548 (1977)
189. Hirata, F., Ohnishi, T., Hayaishi, O.: J. Biol. Chem. *252*, 4637 (1977)
190. Brady, F. O.: FEBS Letters *57*, 237 (1975)

191. Hirata, F., Shimizu, T., Yoshida, R., Ohnishi, T., Fujiwara, M., Hayaishi, O., in: Iron and copper proteins. Yasunobu, K. T., Mower, H. F., Hayaishi, O. (eds.). Adv. Exp. Med. Biol. Vol. 74, New York: Plenum Press, 1976, p. 127.

192. Theorell, H., Holman, R. T., Åkeson, A.: Acta Chem. Scand. *1*, 571 (1947)

193. Tappel, A. L., in: The enzymes. Vol. 8. Boyer, P. D., Lardy, H., Myrback, K. (eds.). New York: Academic Press, 1963, p. 275

194. Oka, T., Simpson, F. J.: Biochem. Biophys. Res. Commun. *43*, 1 (1971)

195. Matsuura, T., Matsushima, H., Sakamoto, H.: J. Amer. Chem. Soc. *89*, 6370 (1967)

196. Krishnamury, H. G., Simpson, F. J.: J. Biol. Chem. *245*, 1467 (1970)

197. Simpson, F. J., Narasimhachari, N., Westlake, D. W. S.: Can. J. Microbiol. *9*, 15 (1963)

198. Oka, T., Simpson, F. J., Child, J. J., Mills, S. C.: Can.J. Microbiol. *17*, 111 (1971)

199. Kido, T., Soda, K., Suzuki, T., Asada, K.: J. Biol. Chem. *251*, 6994 (1976)

200. Hayaishi, O.: Pharm. Rev. *18*, 71 (1966)

201. Kobayashi, S., Kuno, S., Itada, N., Hayaishi, O.: Biochem. Biophys. Res. Commun. *16*, 556 (1964)

202. Sreeleela, N. S., SubbaRao, P. V., Premkumar, R., Vaidyanathan, C. S.: J. Biol. Chem. *244*, 2293 (1969)

203. Gibson, D. T., Koch, J. R., Kallio, R. E.: Biochemistry *7*, 2653 (1968)

204. Gibson, D. T., Cardini, G. E., Maseles, F. C., Kallio, R. E.: Biochemistry *9*, 1631 (1970)

205. Reiner, A. M.: J. Bacteriol. *108*, 89 (1971)

206. Reiner, A. M., Hegeman, G. D.: Biochemistry *10*, 2530 (1971)

207. Hutton, J. J. Jr., Tappel, A. L., Udenfriend, S.: Biochem. Biophys. Res. Commun. *24*, 179 (1966)

208. Hutton, J. J. Jr., Tappel, A. L., Udenfriend, S.: Arch. Biochem. Biophys. *118*, 231 (1967)

209. Kivirikko, K. I., Prockop, D. J.: Arch. Biochem. Biophys. *118*, 611 (1967)

210. Roads, R. E., Udenfriend, S.: Proc. Nat. Acad. Sci. U.S.A. *60*, 1473 (1968)

211. Kivirikko, K. I., Prockop, D. J.: Proc. Nat. Acad. Sci. U.S.A. *57*, 782 (1967)

212. Hausmann, E.: Biochim. Biophys. Acta *133*, 591 (1967)

213. Shaffer, P. M., McCrooky, R. P., Palmatier, R. D., Midgett, R. J., Abbott, M. T.: Biochem. Biophys. Res. Commun. *33*, 806 (1968)

214. Bankel, L., Lindstedt, G., Lindstedt, S.: J. Biol. Chem. *247*, 6128 (1972)

215. Abbott, M. T., Schandl, E. K., Lee, R. F., Parker, T. S., Midgett, R. J.: Biochim. Biophys. Acta *132*, 525 (1967)

216. Holme, E., Lindstedt, G., Lindstedt, S., Tofft, M.: Biochim. Biophys. Acta *212*, 50 (1970)

217. Abbott, M. T., Dragila, T. A., McCrookey, R. P.: Biochim. Biophys. Acta *169*, 1 (1968)

218. Lindblad, B., Lindstedt, G., Lindstedt, S.: J. Amer. Chem. Soc. *92*, 7446 (1970)

219. Abbott, M. T., Udenfriend, S., in: Molecular mechanisms of oxygen activation. Hayaishi, O. (ed.). New York: Academic Press 1974, p. 167

220. Holme, E., Lindstedt, G., Lindstedt, S., Tofft, M.: FEBS Letter *2*, 29 (1968)

221. Goodwin, S., Witkop, B.: J. Amer. Chem. Soc. *79*, 179 (1956)

222. Daly, J. W., Witkop, B.: Angew. Chem., Int. Ed. Engl. *2*, 421 (1963)

223. Hamilton, G. A., in: Progress in bioorganic chemistry. Kaiser, E. T., Kezdy, F. J. (eds.). New York: Wiley, 1971, Vol. I, pp. 83–157

224. Saito, I., Yamane, M., Shimazu, H., Matsuura, T., Chanmann, H. J.: Tetrahedron Lett. *9*, 641–644 (1975)

225. Nakai, C., Nozaki, M., Hayaishi, O., Saito, I., Matsuura, T.: Biochem. Biophys. Res. Commun. *67*, 590 (1975)

226. Schweizer, J., Lattrell, R., Hecker, E.: Experientia *31*, 1267 (1975)

Received March 21, 1978

Author Index Volumes 26–78

The volume numbers are printed in italics

Čársky, P., see Hubač, J.: *75*, 97–164 (1978).

Caubère, P.: Complex Bases and Complex Reducing Agents. New Tools in Organic Synthesis. *73*, 49–124 (1978).

Chandra, P.: Molecular Approaches for Designing Antiviral and Antitumor Compounds. *52*, 99–139 (1974).

Chandra, P., and Wright, G. J.: Tilorone Hydrochloride. The Drug Profile. *72*, 125–148 (1977).

Chapuisat, X., and Jean, Y.: Theoretical Chemical Dynamics: A Tool in Organic Chemistry. *68*, 1–57 (1976).

Cherry, W. R.. see Epiotis, N. D.: *70*, 1–242 (1977).

Chini, P., and Heaton, B. T.: Tetranuclear Carbonyl Clusters. *71*, 1–70 (1977).

Christian, G. D.: Atomic Absorption Spectroscopy for the Determination of Elements in Medical Biological Samples. *26*, 77–112 (1972).

Clark, G. C., see Wasserman, H. H.: *47*, 73–156 (1974).

Clerc, T., and Erni, F.: Identification of Organic Compounds by Computer-Aided Interpretation of Spectra. *39*, 91–107 (1973).

Clever, H.: Der Analysenautomat DSA-560. *29*, 29–43 (1972).

Connor, J. A.: Thermochemical Studies of Organo-Transition Metal Carbonyls and Related Compounds. *71*, 71–110 (1977).

Connors, T. A.: Alkylating Agents. *52*, 141–171 (1974).

Craig, D. P., and Mellor, D. P.: Discriminating Interactions Between Chiral Molecules. *63*, 1–48 (1976).

Cram, D. J., and Cram, J. M.: Stereochemical Reaction Cycles. *31*, 1–43 (1972).

Cresp, T. M., see Sargent, M. V.: *57*, 111–143 (1975).

Crockett, G. C., see Koch, T. H.: *75*, 65–95 (1978).

Dauben, W. G., Lodder, G., and Ipaktschi, J.: Photochemistry of β, γ-unsaturated Ketones. *54*, 73–114 (1974).

DeClercq, E.: Synthetic Interferon Inducers. *52*, 173–198 (1974).

Degens, E. T.: Molecular Mechanisms on Carbonate, Phosphate, and Silica Deposition in the Living Cell. *64*, 1–112 (1976).

Delfino, A. B., and Buchs, A.: Mass Spectra and Computers. *39*, 109–137 (1973).

DeMaine, A. D., see Butler, R. S.: *58*, 39–72 (1975).

DePuy, C. H.: Stereochemistry and Reactivity in Cyclopropane Ring-Cleavage by Electrophiles. *40*, 73–101 (1973).

Devaquet, A.: Quantum-Mechanical Calculations of the Potential Energy Surface of Triplet States. *54*, 1–71 (1974).

Dimroth, K.: Delocalized Phosphorus-Carbon Double Bonds. Phosphamethincyanines, λ^3-Phosphorins and λ^5-Phosphorins. *38*, 1–150 (1973).

Döpp, D.: Reactions of Aromatic Nitro Compounds *via* Excited Triplet States. *55*, 49–85 (1975).

Dougherty, R. C.: The Relationship Between Mass Spectrometric, Thermolytic and Photolytic Reactivity. *45*, 93–138 (1974).

Dryhurst, G.: Electrochemical Oxidation of Biologically-Important Purines at the Pyrolytic Graphite Electrode. Relationship to the Biological Oxidation of Purines. *34*, 47–85 (1972).

Durr, H.: Reactivity of Cycloalkene-carbenes. *40*, 103–142 (1973).

Dürr, H.: Triplet-Intermediates from Diazo-Compounds (Carbenes). *55*, 87–135 (1975).

Dürr, H., and Kober, H.: Triplet States from Azides. *66*, 89–114 (1976).

Durr, H., and Ruge, B.: Triplet States from Azo Compounds. *66*, 53–87 (1976).

Dugundji, J., and Ugi, I.: An Algebraic Model of Constitutional Chemistry as a Basis for Chemical Computer Programs. *39*, 19–64 (1973).

Dugundji, J., Kopp, R., Marquarding, D., and Ugi, I.: *75*, 165–180 (1978).

Eglinton, G., Maxwell, J. R., and Pillinger, C. T.: Carbon Chemistry of the Apollo Lunar Samples. *44*, 83–113 (1974).

Eicher, T., and Weber, J. L.: Structure and Reactivity of Cyclopropenones and Triafulvenes. *57*, 1–109 (1975).

Volume 12
Immobilized Enzymes II

1979. Approx. 180 pages
ISBN 3-540-09262-5

Contents:
H. M. Koplove, Ch. L. Cooney: Enzyme Production During Transient Growth – The Reorganization of Protein Synthesis. – *R. D. Schmid:* Stabilized Soluble Enzymes. – *S. S. Wang, C. King:* The Use of Coenzymes in Biochemical Reactors. – *C. Wandrey, E. Flaschel:* Process Development and Economical Aspects in Enzyme Engineering: Acylase-L-Methionine System. – *D. J. Graves, Yun-Tai Wu:* The Rational Design of Affinity Chromatography Separation Processes.

Volume 7
Biotechnology

1977. 112 figures, 14 tables. V, 150 pages
ISBN 3-540-08397-9

Contents:
K. Schügerl, J. Lücke, U. Oels: Bubble Column Bioreactors. Tower Bioreactors without Mechanical Agitation. – *R. Acton, J. D. Lynn:* Description and Operation of a Large-Scale, Mammalian Cell, Suspensio Culture Facility. – *S. Aiba, M. Okabe:* A Complementary Approach to Scale-Up Simulation and Optimization of Microbial Processes. – *L. Kjaergaard:* The Redox Potential: It Use and Control in Biotechnology.

Volume 8
Mass Transfer in Biotechnology

1978. 95 figures. V, 151 pages
ISBN 3-540-08557-2

Contents:
M. Charles: Technical Aspects of the Rheological Properties of Microbial Cultures. – *K. Schügerl, J. Lücke, J. Lehmann, F. Wagner:* Application of Tower Bioreactors in Cell Mass Production. – *M. Zlokarnik:* Sorption Characteristics for Gas-Liquid Contacting in Mixing Vessels.

Volume 9
Microbial Processes

1978. 69 figures, 15 tables. V, 144 pages
ISBN 3-540-08606-4

Contents:
V. L. Yarovenko: Theory and Practice of Continuous Cultivation of Microorganisms in Industrial Alcoholic Processes. – *Y. Miura:* Mechanism of Liquid Hydrocarbon Uptake by Microorganisms and Growth Kinetics. – *J. E. Zajic, N. Kosaric, J. D. Brosseau:* Microbial Production of Hydrogen. – *T. Enatsu, A. Shinmyo:* In vitro Synthesis of Enzymes. Physiological Aspects of Microbial Enzyme Production.

Volume 10
Immobilized Enzymes I

1978. 48 figures, 14 tables. VII, 177 pages
ISBN 3-540-08975-6

Contents:
W. H. Pitcher: Design and Operation of Immobilized Enzyme Reactors. – *S. A. Barker:* Biotechnology of Immobilized Multienzyme Systems. – *R. A. Messing:* Carriers for Immobilized Biologically Active Systems. – *P. Brodelius:* Industrial Applications of Immobilized Biocatalysts. – *B. Solomon:* Starch Hydrolysis by Immobilized Enzymers.

Volume 11
Microbiology, Theory and Application

1979. 76 figures. Approx. 200 pages
ISBN 3-540-08990-X

Contents:
D. Ramkrishna: Statistical Models of Cell Populations. – *Sh. Nagai:* Mass and Energy Balances for Microbial Growth Kinetics. – *M. Moo-Young, J. M. Scharer:* Methane Generation by Anaerobic Digestion of Cellulose-Containing Wastes. – *B. Metz, N. W. F. Kossen, J. C. Van Suijdam:* The Rheology of Mould Suspensions. – *M. Zlokarnik:* Scale-Up of Surface Aerotors for Waste Water Treatment.

Springer-Verlag
Berlin
Heidelberg
New York

Structure and Bonding

Editors:
J. D. Dunitz, P. Hemmerich, J. A. Ibers,
C. K. Jørgensen, J. B. Neilands, R. S. Nyholm,
D. Reinen, R. J. P. Williams

Springer-Verlag
Berlin
Heidelberg
New York